Spatial Theories of Education

Routledge Research in Education

1. Learning Communities in Education
Edited by John Retallick, Barry Cocklin and Kennece Coombe

2. Teachers and the State
International Perspectives
Mike Bottery and Nigel Wright

3. Education and Psychology in Interaction
Working with Uncertainty in Inter-Connected Fields
Brahm Norwich

4. Education, Social Justice and Inter-Agency Working
Joined up or Fractured Policy?
Sheila Riddell and Lyn Tett

5. Markets for Schooling
An Economic Analysis
Nick Adnett and Peter Davies

6. The Future of Physical Education
Building a New Pedagogy
Edited by Anthony Laker

7. Migration, Education and Change
Edited by Sigrid Luchtenberg

8. Manufacturing Citizenship
Education and Nationalism in Europe, South Asia and China
Edited by Véronique Bénéï

9. Spatial Theories of Education
Policy and Geography Matters
Edited by Kalervo N. Gulson and Colin Symes

Spatial Theories of Education

Policy and Geography Matters

Edited by
Kalervo N. Gulson and
Colin Symes

Routledge
Taylor & Francis Group
New York London

Routledge
Taylor & Francis Group
270 Madison Avenue
New York, NY 10016

Routledge
Taylor & Francis Group
2 Park Square
Milton Park, Abingdon
Oxon OX14 4RN

Routledge is an imprint of Taylor & Francis Group, an Informa business

Printed in the United States of America on acid-free paper
10 9 8 7 6 5 4 3 2 1

International Standard Book Number-10: 0-415-40395-2 (Hardcover)
International Standard Book Number-13: 978-0-415-40395-5 (Hardcover)

Library of Congress Cataloging-in-Publication Data

Spatial theories of education : policy, geography and theory / edited by Kalervo N. Gulson and Colin Symes.
p. cm. -- (Routledge research in education ; 9)
Includes bibliographical references and index.
ISBN-13: 978-0-415-40395-5 (hardback : alk. paper)
1. Educational sociology. 2. Education--Philosophy. 3. Geography--Philosophy. I. Gulson, Kalervo N. II. Symes, Colin, 1945-

LC189.S693 2007
306.43--dc22 2006033876

Visit the Taylor & Francis Web site at
http://www.taylorandfrancis.com

and the Routledge Web site at
http://www.routledge.com

Contents

Figures and Tables

Figures

Tables

Acknowledgments

Books as they are being written seem to have personalities and lives of their own. They grow from small beginnings, from the seed of an idea (and here Deborah Youdell played a crucial role as the person who first encouraged Kalervo to consider the links between space and policy) to something that is almost too large to manage. And, on their way to their final forms, books can reveal difficult, obstreperous personalities, though most of the time they are a joy to be with. This one was no different and during its 'creation' manifested both difficult and joyful sides. In helping us to get through the difficult periods and to appreciate the joyful ones, there are a number of people we would like to thank.

First, we would like to thank the other authors in this book (many of whom were participants in two symposia on space and education at the American Educational Research Association annual meetings in 2005 and 2006), who wrote their chapters in good time, and responded to our criticisms with sensitivity and patience. We literally could not have produced this book without their forbearance and cooperation. Second, there are the various behind-the-scenes people who played a significant role in bringing this book into existence and giving it a more pleasing shape and feel. There is our commissioning editor Terry Clague at Routledge; there are our immediate colleagues at Charles Sturt and Macquarie universities who have been in various ways inconvenienced by our working on the book. We would also like to thank Shyam Howlin who helped to iron out the word processing problems of working across multiple word processing formats. We would also like to acknowledge our gratitude to Lorraine Shannon who helped us with the copy-editing. We were able to employ Lorraine as a result of a grant from Macquarie University's School of Education. Gratitude also goes to the Macquarie University's Australian Centre for Education Studies for extending an invitation to Kalervo Gulson to become a Visiting Fellow in the Centre during the first semester of 2006 and to the School of Teacher Education, Charles Sturt University, for providing Kalervo with the opportunity to undertake this fellowship.

On the home front, we would like to apologize to our partners for having to suffer the effects of the work-at-home syndrome, for the book intruding on our time together, during evenings and at weekends, and promise that next time it won't happen!

1 Knowing one's place
Educational theory, policy, and the spatial turn

Kalervo N. Gulson and Colin Symes

> geography…now looks set to become the sexiest academic subject of all.
>
> (Eagleton 1997)

Educational theory moves by fits and starts, and often takes its developmental cues from other disciplines. It is not that educationalists are incapable of generating original thought that might lead to application elsewhere in the disciplinary spectrum, but more that they tend to be the followers of broad epistemological trends or 'turns' rather than their creators. In the latter parts of the twentieth century there was a resurgence of 'space' in social theory, a resurgence, or 'turn', that education has yet to address, at least in any concerted way. The word 'concerted' suggests that the turn might have begun to occur, albeit in isolated pockets, in and around education, and that there are already incipient theories of space pertaining to matters educational that demand wider attention and systematic analysis. And one of the objectives of this chapter, and book, is to demonstrate that this indeed is the case: that 'space' is luxuriant with ideas, concepts, and theories, which, if employed, could result in education taking a new turn. As such the conceptual framework for this book is necessarily an eclectic one that exemplifies the diverse and contested notions of space and their application in education. Like any text this book excludes and includes, and while we have been at pains not to advocate one position, many of the chapters do focus on a social constructivist view of space, and, in some ways, there is a reliance on the work of the contemporary spatial 'canon' (C. Taylor, personal communication)—composing the work of Henri Lefebvre, Doreen Massey, and Edward Soja.

The contributors to this book recognize that spatial theories are not restricted to geography, their traditional 'home', but travel through and between social theory and are 'implicated in myriad topographies of power and knowledge' (Gregory 1994: 11). As its title suggests this book is concerned with theory, with the adoption of 'theory' as possibility, working to unsettle, to destabilize, to shift assumptions in educational policy studies. The deployment of theory by the authors in this collection rests comfortably with what Stephen Ball calls the 'necessity and violence of theory' (Ball

2006). This is also congruent with the central premise of this book: that drawing on theories of space contributes in significant and important ways to subtle and more sophisticated understandings of the competing rationalities underlying educational policy change, social inequality, and cultural practices. Therefore, examining education policy from a spatial perspective is not about creating 'new' problems as such, but rather it is about providing explanatory frameworks that, perhaps, disrupt understandings in, and posit new possibilities for, 'mainstream' education policy studies.

The immediate and initial possibilities of this can be seen in terms of the everyday language of space. This chapter's title is not atypical of how the language of space, perhaps because it is linked so palpably to experience, is imbued with the sense of inclusion and exclusion as individuals and groups move through, in, and out of communities. There is a sense that knowing one's place has a powerful sociological resonance; stories and narratives mediate the way space is apprehended and comprehended. Because spatial manoeuvre is such a key part of quotidian life, much of our vocabulary is concerned with specifying the fundamental ordinates of space, with communicating information about position, direction, and movement. Words such as 'here' and 'there' are critical to establishing proximity and immediacy, and in conjunction with modifiers such as 'closer', 'further', and 'nearer' provide a way of calibrating and mapping the spatial experience. Space also provides metaphors for understanding the nature of society and social experience (see Bourdieu 1999).

The spatial turn tends to emphasize the transient and social nature of space, that space is a construct *not* a given. Here it is usual to contrast space with place, the more nominal sense of space. For example, places generally have names; they figure on maps, have boundaries and parameters—there is an element of fixity pertaining to them. Place lends itself to more 'objective', scientific accounts of space—even though what constitutes a place is itself a construct, subject to myriad judgements as to where its boundaries and populations begin and end (for debates on space and place see Agnew 2005; Casey 1996; Harvey 1993; Massey 1993b). Space is more generic, more amorphous and porous, hard to pin down. It is more subjective, more quotidian—not the object of parenthesizing and naming in quite the same way place is. The fact that it is usual to refer to spatial practices is also revealing: places do not have practices in quite the same way. Space in this sense is more of a verb than a noun.

However, there are also problems with these understandings that we recognize, and we have attempted to create a cautious, and careful, collection that is attendant to Gregory's call for us to 'interrogate…"common-sense" understandings' (1994: 12). For in the early 1990s, geographers Neil Smith and Cindi Katz (1993) warned, *pace* Gregory, of the dangers of the proliferation of spatial discourses in the social sciences. They suggested that while 'space' and 'place' had escaped their traditional disciplinary confines, something we will take up below, the histories of scholarship and significant debates about their salient meanings had been largely left behind when they migrated; they had effectively been lost in translation. Apropos this loss, Smith and Katz (1993) argue that the ideas of space and place are more often used by social scientists as metaphors, rather than as complex theorizations of material and symbolic life. A spatial metaphor 'can only suggest and stimulate further thought: it cannot provide an answer. But, too often, it is assumed that this is what it does' (Thrift and Whatmore 2004: 8).

While these critiques might be simply dismissed as the backlash of a field of study losing its semantic monopoly, and as a modernist ploy to fix meaning (P. Thomson, personal communication, 2006), the proposition that there might be important debates and traditions of theorizing that are of potential interest to social science researchers in general, cannot be so easily put aside. Thus, this book takes the position that space and place are integral, yet under-examined and under-theorized, components of policy studies, development and critique.

In this chapter we briefly set out the key 'trajectories' of space in social theory. We also examine what happens when spatial theories 'escape' traditional disciplinary confines and take on, using Lefebvre's term, 'transdisciplinary' elements. As Soja defines it 'transdisciplinarity' is 'not being the privileged turf of such specialized fields as History, Sociology, and Geography, but spanning all interpretive perspectives' (Soja 1996: 6). Importantly, this raises the issue as to what extent education is *education* any longer when it takes up the spatial dimensions of its field? The contributors to this book, while not wrestling directly with this question, are nonetheless expanding the field of education into new territory, an expansion which the last part of the chapter 'maps'.

FROM EUCLID TO SOJA

It is of note that while educational history is a long established field, with a celebrated body of literature and scholarship that has contributed to understanding the way educational systems have developed, educational geography—its spatial equivalent—remains relatively underdeveloped and 'unnamed' though a few have identified themselves as practising it (e.g. Marsden 1977, 1987).[1] The absence of a well-defined field devoted to examining the 'spatial' questions and dimensions of education may be due to a long tradition of history assuming an ascendancy over geography in social theory. At the core of this tendency is the issue of mutability, that time is more amenable to change, specifically political change, than space, which was held to be incorrigible and resistant to transformation. Nonetheless, as Mike Crang and Nigel Thrift (2000) argue, as do others such as Massey (1993a):

> space is exceedingly hard to write about shorn of its relation to time. Though part of the reason for the turn to space in many disciplines has been a drive to move away from the tyrannies of historicism and developmentalism, the fact remains that space without time is as improbable as time without space. (Crang and Thrift 2000: 1)

Arguably, Platonist tendencies evident in the scientific and mathematical conceptualization of space held back the appreciation of space in social terms. The notion of a positivist 'spatial science,' particularly in the 1950s and 1960s, looked to the development of theory on the basis of statistical analysis; a 'quantitative revolution' in geography premised on the ideas of 'abstract space' that 'conceive[s] of space as a surface on which the relationships between (measurable) things were played out' (Hubbard, Kitchin, and Valentine 2004: 4). In a similar vein Massey (1992, 1994) has noted how 'space' in particular was very often written about as if it were an empty vessel within which action took place, or as an effect of social, political, and economic relations. These codifications were characterized by abstractionist impulses that necessitated space being uprooted from its contexts and which, arguably, were not useful for the humanities and social sciences where it is what individuals and societies do with space that 'counts'. Quantitative approaches have also been criticised for not engaging with other perspectives such as feminist geography (e.g. Falconer Al-Hindi 2001). Nonetheless, some have argued against an either/or approach to space, for as Trevor Barnes and Matthew Hannah (2001) contend, in relation to statistics: 'how and why they are constructed, by whom and about whom, how they are used and mobilised, and to what ends and for whose interests, are critical social scientific questions for critical social scientists' (pp. 379–80).

Indeed, there developed the view that mono-disciplinary approaches to space, be they of a sociological, historical, or geographical kind, might be insufficient and that a 'transdisciplinary perspective' might provide more insights. Urban sociologists and geographers interested in the connection between space and capitalism challenged the 'quantitative revolution' because it downplayed the political relationship between space and social relations. What is certainly a feature of these moves after the 'quantitative revolution' is the shift towards '*space as process* and *in process* (that is space and time combined in becoming)' (Crang and Thrift 2000: 3, original emphasis).

In this book the use of space is mostly concerned with making sense of materiality—in many ways this ascribes to Crang and Thrift's (2000) contention that despite different adoptions of space across a variety of disciplines, 'in all disciplines, space is a representational strategy' (p. 1). Yet this too has its problems. As Neil Smith writes:

> More generally, a naïve social constructionism which, suspicious of materiality, treats space first and foremost as representational or con-

> ceptual, turns out to be paradoxically conservative; it fails to unearth conceptions of space that differ significantly from traditional Enlightenment treatments of space...which in many respects represented an elaboration and universalisation of the presumptions of Euclid's geometry. (Smith 2004: 17)

Smith argues that it was Henri Lefebvre's book *La production l'espace* (1974, translated into English as *The Production of Space*, in 1991)[2] that provided the opportunity to systematically address these problems. This work has proved enormously influential in the last decade or so and is crystallized around the idea that, in relation to capitalism, space is socially produced, engineered, and constructed, and that social relations are *always* constituted relative to space (Lefebvre 1991). Lefebvre is certainly a presence in this collection with many of the authors referring to his work on perceived, conceived, and lived space. Lefebvre's contemporary interlocutor, Edward Soja, with his Lefebvrian inspired spatial trialectic of 'spatiality, historicity, and sociality' (see Soja 1996; Chapter 3) is also a key reference point for many of the chapters. Many of the contributors also work with post-modern and post-structural ideas in productive tension with neo-Marxism (M. Peters, personal communication). These authors are working with, and within, 'lived space' as a realm of the imagination and as an incitement of the possible.

PARADIGMS AND GENEALOGIES OF THE SPATIAL TURN

In many ways the spatial trajectories in education have mirrored those of other disciplines in borrowing, appropriating, and employing the spatial theories of geography. We can see that much of the existing literature using, or referring to, space is 'transdisciplinary', encompassing such fields as art history and economics. Thrift and Whatmore (2004) assert that 'whereas culture has been a shared currency that has kept geographers in conversation with others for a long time now, it is only more recently that space has begun to take on a similar iconic status, spinning out of geography through the intellectual curiosities and investments of many other disciplines' (p. 2). The move towards drawing on these spatial ideas that have predominantly been played out in human (cultural and social) geography, is also part of, in terms of multiple trajectories of inquiry, what might be seen as a particularly post-modern exercise (e.g. Mourad 1997), one that, perhaps somewhat ironically, defies the modernist project of rigid classification and disciplinary policing.

So, for example, we can see some of these aspects in the particular instances of cultural geography and sociology of education. One of the key aspects of cultural geography has been its affinity with the 'extra' disciplinary—a resistance to being 'disciplined' in any strict sense (Thrift and

Whatmore 2004). These endeavours draw on a number of 'in-between' disciplinary zones such as post-colonialism and feminist studies, which have also been a feature of recent sociology of education (see Ball 2004). Indeed, it is perhaps pertinent to see how ideas have taken flight within disciplines, as well as between them, with a number of special issues of journals in both education policy studies and human geography devoted to exploring theoretical contributions such as post-structuralism.[3] In a sense what all the contributors in this book are doing is exploring the realms of possibility that follow from the blurring or jumping of boundaries. Joe Kincheloe (2001) claims that:

> what we refer to as the *traditional disciplines* in the first decade of the 21st century are anything but fixed, uniform, and monolithic structures.... We occupy a scholarly world with faded disciplinary boundary lines. (Kincheloe 2001: 683, origianl emphasis)

Within the imbricated nature of the education 'field' certain approaches fall out of favour, only to come back into favour at some stage; history one moment, geography, when the moment is right, the next. These are more than shifts in 'fashion' and speak to a mutually constitutive relationship between objects of study and research approaches in explorations of educational policy processes and practices. Yet in exploring these theoretical 'lines of flight' in terms of disciplinary categories, in the sense, perhaps, of 'lines of flight or of deterritorialization' (Deleuze and Guattari 1987: 32), what may also matter is the way in which we present these lines. For example, the way in which we have presented the 'spatial shifts' above emphasizes their linear development, with a similarity to the paradigmatic shifts between periods of revolutionary and normal science (Kuhn 1962). However, there are problems with this type of presentation, for as Hubbard et al. (2004) suggest, '[t]he danger of a paradigmatic approach to understanding the geographical tradition is that it creates a linear narrative that suggests that spatial thought has developed through unified (and generational) paradigms when in reality consensus has seldom been complete or stable' (11). Similarly, Ball, when focussing on a variety of turning points, or 'disputational moments' in sociology of education, such as feminism and anti-essentialism claims that:

> I am not intending to suggest that what has taken place are sort of revolutionary paradigm shifts that Kuhn identifies in the natural sciences. These are not points of thorough-going conversation across the whole discipline, rather they contribute to the proliferation and sedimentation of perspectives and standpoints, the multiple traditions within the sociology of education. (Ball 2004: 3)

Alternatively, genealogical approaches focus on how ideas emerge and are adopted, even adapted, how particular forms of knowledge are 'situated' (Haraway 1999), or emerge as matters of concern because of particular conditions or blending of conditions, e.g. globalization and neo-colonialism. Thus, 'spatial thought is not developed in a vacuum, but is rather constructed by individuals (and individuals collaborating) and situated within their own personal and political beliefs, the culture of academia, and institutional and social structures' (Hubbard et al. 2004: 11). Here it is significant to recognize that each of the contributors to this collection comes to 'space' in particular ways; all are 'situated' spatially within variegated contexts (e.g. institutional, national, ideological), and 'situated' temporally, within a particular moment of education policy studies. This also poses challenges for ways of understanding the location of ideas and drawing disciplinary boundaries. Donna Haraway talks of this type of activity in terms of:

> bodies as objects of knowledge are material-semiotic generative nodes. Their *boundaries* materialize in social interaction. Boundaries are drawn by mapping practices.... But boundaries shift from within; boundaries are very tricky. What boundaries provisionally contain remains generative, productive of meanings and bodies. Siting (sighting) boundaries is a risky practice. (Haraway 1999: 185, original emphasis)

'Risky practice' is also a useful way of thinking about some of the possible problems with this book—and with the potential and possibilities of space in policy studies work. Just what does it mean for educational researchers to claim that they are using space? While the allure of theory is that it appears able to work between, across, within, and 'in-between' disciplinary boundaries, alternatively, instead of destabilizing boundaries, and being generative, there is always a danger of creating a 'fetishism of space' (Smith 2004), and through imprecision disciplinary rigidity may inadvertently be reinscribed and 'space' reclaimed by geography. To that end, we suggest there is a need to seriously address Neil Smith's challenge around space:

> [t]here is a crucial question of the extent to which this 'spatial turn' has been more than skin deep.... Put most crudely, perhaps, why space? Why *should* our analysis of social difference and political possibility be rewritten in the language of space? (Smith 2004: 13, original emphasis)

Some answers, therefore, to why 'space' in education may lie in the objects of study with which policy studies and the sociology of education have been concerned.

EDUCATION AND SPACE

Spatial analyses of various kinds and forms have been undertaken for at least a hundred years, if not more. There has been long tradition of cultural analysis stemming back to the Arcades' Project of Walter Benjamin who, in his turn, took his cues from Baudelaire, recognizing the cornerstone role played by the city in the generation of modernity and modernist subjectivity (1999). Benjamin's enormously influential work prompted studies of contemporary cities. Schools and universities, like other cultural institutions, are important sites in the constitution of the modern city, though the significance of these sites is often downplayed and has not received the same attention as shopping malls and theme parks (e.g. Shields 1991; Zukin 1998).

Educational space manifests itself in myriad ways; one of the most obvious is school architecture, comprising spaces specifically designed with the processes of teaching and learning in mind. Nonetheless, there has been a tendency to treat school architecture as a background phenomenon that was not integrated into the power of pedagogy, or, if it was, then it was seen as an innocuous force, generally for the educational good. Further, what studies of school architecture there were concentrated on the 'look' of the school, as exemplifying a particular style of educational aesthetic (e.g. Seaborne 1971; Seaborne and Lowe 1977). Rarely did scholars of school architecture consider that there was a life inside the school, that there was more to schools than mere architectural plans and floor layouts, and that the school, as a result of the practices that went on therein, was constantly subject to re-design and re-formation. Whatever notion of space adhered to the analysis of school architecture, it tended to be a fixed and stationary one, which took no account of the fact that educational spaces are fluid and ephemeral, ever being re-written and re-inscribed as each pedagogic moment is transformed into another and as they are acted out in time, with a 'determined repertoire of behaviour' (Benito 2003: 53).

And only very occasionally did they attempt to evaluate the policy context that brought into existence, and gave a 'face' to certain school designs and discouraged others (see Saint 1987). The work of Michel Foucault (e.g. 1977) helped to change these tendencies and led to a flood of literature analyzing buildings as places of surveillance and regulation (King 1980; Lawn and Grosvenor 2005; Markus 1993; McGregor 2004; Symes 1996, 1999). School buildings, and architecture (e.g. Ellsworth 2005), far from being the innocent backdrop to pedagogy were recast as central to the administration of the school population, providing the fabric of a disciplinary technology that, through the spectre of unremitting inspection and surveillance, enabled it to be normalized and classified on a day-to-day basis.

Space has also been instantiated in other elements of educational life,[4] particularly with the emergence of a neoliberalist agenda that has brought into spotlight the differentiating spatial effects of unequal provision. Work in the UK, where the educational effects of this agenda have had their most pronounced and pervasive impact, has drawn attention to the degree to which the geography of schooling is part and parcel of the strategy of 'choice' (Ball 2003; Ball, Bowe, and Gewirtz 1995; Taylor 2001; Vincent and Martin 2002). This includes work covering areas such as educational policy change around questions of identity (Gulson 2005; Vincent, Ball, and Kemp 2004), including 'race' (Ball, Maguire, and Macrae 1998; Gulson 2006). Arguably, and this is not to condone it, the neo-liberalist policy framework has forced educationalists to consider the spatial facets inherent in marketized approaches to educational provision.

Another, and somewhat related, development that has forced space into the educational spotlight is globalization—though to what extent globalization can be understood as a discrete phenomenon, separate from the neoliberal agenda is, at times, a point of contention. And certainly a spatial aspect to globalization is around the notion of policy travelling and borrowing (e.g. Ball 1998; Dale 1999; Henry, Lingard, Rizvi, and Taylor 2001; Levin 1998), in a sense, the extension and compression of policy processes and practices.

Additionally, space has been drawn on in work around literacy (Leander and Sheehy 2004), curriculum (Edwards and Usher 2003) and critical pedagogy (Callejo Pérez, Fain, and Slater 2004). In addition work on the geography of disability (e.g. Hall 1999) has also been considered in terms of disability and education (e.g. Armstrong 2003). Yet there are domains of education in which space is a preoccupation of pedagogy that have not been hitherto examined. These include the whole question of spatial representation and the nature of its visualization in maps, paintings, and diagrams (see Massey 2005: 1–4) and which lie at the heart of much teaching in mathematics, geography, and science; then there is the matter of the utilisation of imaginary literary spaces, or 'spaces of writing' (Crang and Thrift 2000: 22), in novels and poetry. And so on.

'MAPPING' THE COLLECTION

What characterizes this book is not that there was any direct collaboration between its contributors prior to writing their chapters, but rather the way that the produced chapters result in a series of conversations, dealing with a number of themes, from the impact of globalization on the movement of students to the development of education policy within inner cities. Although these themes might initially seem discrepant, what unites them are the possibilities provided by spatial analysis; that current theorizing of space helps edify a variety of educational problems, on, and across, a range of scales. Notwithstanding the reach and diversity of the issues covered within the book, the ambits of the ensuing thirteen chapters fall within three distinct domains of concern. The initial five provide strong theoretical frameworks for understanding the spatial dimensions of policy; the next four adopt a case study approach, examining at a local school, suburban, or city 'scale' the mutual constitution of space and inequality; and the final four deal predominantly with issues of scale, mobility, and identity to investigate issues of globalization, resistance, and changes to Western universities.

What characterizes the first five chapters is a concern with foregrounding theoretical premises and using empirical data analysis as illustrative examples. These chapters show the diverse range of theoretical possibilities open to the educational policy researcher when using space, these possibilities working with a number of diverging epistemological positions from the broad concerns of a constructivist scalar politics, to the resurgence of quantitative geographies as represented by Geographical Information Systems (GIS).

Leading off this collection, Claudia Hanson Theim outlines a politics of scale and place to posit the types of analysis supported by what she calls a 'constructivist geographical imagination'. Applying this imagination to an examination of homeschooling in the United States, Thiem concludes that, through the mobilization of a politics of scale, the home schooling movement has demonstrated a capacity to create a 'spatial networking strategy' to 'jump scales' (from the family to the state), and, as an outcome, has constructed the 'home-school' as a legitimate space of resistance to the state.

Kalervo Gulson utilizes a post-structural spatial analytic, with a focus on relational space, to examine the political mobilization of space discourses as part of educational policy change in the inner city of Sydney. Gulson identifies the community consultation process, which took place under the aegis of a parliamentary inquiry, as a case study of spatialized and historicized knowledge/power. His chapter attempts to exemplify the explanatory force of post-structural geography for educational policy sociology.

Identifying the possibilities of thinking about space and education differently is Bill Green and Will Letts's central aim, specifically through what they call 'spatial-difference dynamics' and the connection to the educational politics of rural schooling in New South Wales (NSW). Their chapter draws on various 'lines of flight' within social theory such as following Soja (1996) to posit a 'trialectic' of rurality/education, space, and equity. Green and Letts explore this 'trialectic' through a brief overview of the way a spatial dimension was incorporated into the state's vision of (rural) education and, as their illustrative case study of the NSW region of Dubbo shows, how this was subsequently abandoned.

Geographical Information Systems (GIS) are now employed in social research and have, as Chris Taylor argues, application in education, particularly in mapping the geography of school choice. Taylor also links with the traditions of 'quantitative' geography and predictive policy analyses in the use and manipulation of post-code data to predict the possible impacts on school enrolment from reorganisation or from changes to school admissions policies. Whilst Taylor's chapter focuses on the issues of school choice and education marketization in the United Kingdom, he also points to the other possibilities in educational research that may emerge from using GIS as a form of spatial analysis.

Exploring the potential of space and place in the analyses of the processes and practices of exclusion and inclusion in the United Kingdom is Felicity Armstrong's project. Soja's (1996) notion of 'Thirdspace' is employed as a stimulus, as a way of reworking studies of disability and exclusion, with an aim of thinking 'creatively' about how people, and children specifically, are 'spaced out' of 'mainstream settings'. What Armstrong challenges us to consider is whether 'uni-discplinary' ideas are, in themselves, adequate in understanding processes and practices of marginalization and segregation.

The middle four chapters of the collection have a much stronger emphasis on particular case studies as a way of demonstrating the creative possibilities of space and place in educational policy studies. These studies, whilst concerned with carefully identifying the spatial premises of the work, may be considered 'local' in focus, with the objects of analysis ranging from the 'school' to the 'city', with issues of, and interconnections between, identity and exclusion being obvious.

In a similar vein to Felicity Armstrong, Pat Thomson brings a spatial consideration to the issue of the visible and invisible 'geographies of exclusion' within school settings, namely the evaluation of a 'multi-site' project in the United Kingdom that aims to provide 'excluded' students with a 'second chance' at a new school. Drawing on notions of perceived and lived spaces, what is central to Thomson's chapter is not only that students are made vulnerable by processes and practices of exclusion, but that there are spatial alternatives to this exclusion, namely resistance, changes in place, and the construction of counter-public spaces. Through these spatial alternatives there are possibilities for students to build identity and agency.

Jo-Anne Dillabough, Jacqueline Kennelly, and Eugenia Wang outline a 'hermeneutic and praxiological' framework to understanding the interplay of gender, class, youth culture, and school and urban spaces. The central focus is on enactments of peer rivalry in an urban area in southern Ontario, Canada. Dillabough et al. are interested in the ways that 'young people's phenomenologies of meaning' are connected to the cultural aspects of political and spatial power, social hierarchies, and privilege. The authors demonstrate how spaces are constituted by and are constitutive of, youth sub-cultures, through sites such as schools, particularly school corridors, and an urban area that is suffering from the consequences of neoliberal reforms.

Pauline Lipman employs spatial theories to demonstrate how educational policy change, in this instance Renaissance 2010 in Chicago, intersects with urban changes such as gentrification. It is suggested that the interests of real estate developers, global corporations, and city planners increasingly coincide in the re-engineering of urban environments. Lipman also argues that education policy is implicated in questions of spatialized inequality and identity (e.g. race) in the city. Additionally, education is posited as 'an arena of struggle' in terms of the constitution of urban space; a struggle that involves multiple scales from the 'local' (gentrification) to the 'global' (neoliberalism).

Colin Symes examines the transformation wrought on Sydney when its suburban railway network was established. Utilizing a historical geography approach he argues this produced a segregated city, in which the private schools, represented by various churches, recognized that proximity to the rail system was a distinct positional advantage. An important principle was set in train: that the preparedness to be mobile is a key element of maximizing school choice, which has long remained a practice in Sydney. Symes's chapter demonstrates that spatial problems also become temporal ones, that once established cannot easily be undone.

The last four chapters involve spatial concerns around scale, mobility, and identity. Scale analyses are posited as useful in investigating social change as they construct spatial categories into hierarchical orders (e.g. local/global), and in examining aspects of power relations such as 'domination' and resistance. Mobility studies has been closely associated with the reworking of modernity, such as Zygmunt Bauman's (2000) notion of 'liquid modernity' (see also Urry 2000, 2005). Bauman argues that modernity is increasingly more liquid than solid, transient, in a permanent state of kinesis and that pace forms an important ternary with space and place. This has been fuelled by transport and electronic technologies, which have created unprecedented flows: of information, of commodities, peoples, of cultural exchange, and so on (Appadurai 1996). And here we might look at the 'mobility' of ideas, connected to not only technological and travelling factors but also to the constitution of 'new' types of, and identities for and by, peripatetic academics and students.

Michael Singh, Fazal Rizvi, and Mona Shrestha deal with the international 'trade' in higher education students. Their study employs Lefebvre's (1991) trialectic of perceived, conceived, and lived space to examine the undoing of local, national and international boundaries. The authors interview a cohort of Chinese students, undertaking post-graduate studies at Australian universities, to unpack issues of mobility and identity. The study reveals the ambivalence that the students feel towards their host educator nation but also the transformation of identity that they experience as they come to terms with 'their' Australia.

The European Union (EU) is a new scalar phenomenon, the 'multinational state'. Susan Robertson investigates the EU's education policy as it seeks to advance its economic interests through the knowledge economy. This represents a new space for policy making which, in the European context at least, involves the participation of trans-national corporations working alongside government organizations in 'Public-Private Partnerships'. As Robertson sees it, these partnerships have placed the imperatives of the new economy at the forefront of their concerns. Older notions to do with fostering opportunity that focused on developing social capital are marginal in the race to digitalize EU populations. Robertson suggests that these developments require educational researchers to develop new sets of theoretical tools of analysis, in this instance a focus on how 'space matters'.

Similar to Robertson, Bob Lingard also argues that education policy making is now multi-scalar, and is increasingly, though not exclusively, articulated outside the boundaries of the nation state. Using the spatial aspects of post-colonial theory in conjunction, and sometimes in tension, with the work of Pierre Bourdieu, and deploying the work of Arjun Appadurai, with a focus on globalization from below, Lingard brings a critical focus to the phenomenon of international students in Western universities. Drawing on his experiences of teaching international students on- and offshore, Lingard notes how responses to globalization among his students are differentiated ones: such as globalization as an opportunity to place China at the forefront of international development; and a bottom-up version of globalization that affirms Caribbean culture, and resists the hegemonic features of globalization.

We finish the collection by connecting back to Claudia Hanson Thiem's chapter on the politics of scale. Mario Novelli chronicles worker and community inspired action to resist the privatization of Colombia's state-owned utilities. This is the scalar politics of resistance that challenges the notion that capital is global whilst labour is local. This resistance was led by trade unions who organized through what Novelli calls 'strategic pedagogy'. Employing the spatial theories of Kevin Cox (1998), Novelli points out how, across scales, the Colombian cause was advanced through particularly UK-based trade union and social organizations, and international human rights bodies such as Amnesty.

CONCLUSIONS

This collection makes no claims to provide a comprehensive overview of educational research that employs spatial theories, rather we think of this collection in the same way the authors have employed concepts of space: as a catalyst, providing possibilities for disruption, and a demonstration of the potential directions within, and for, educational policy studies.

The use of spatial theories appears to challenge the notion of what it is to study 'education'. In many ways these are equivalent arguments to those in human geography, and whilst we recognize the danger of simply 'dislocating' ideas the potential of the porous nature of educational research and of human geography is evidenced in the contributions in this collection. The failure to entertain in any full-blooded way, the spatial dynamics and exigencies underpinning education means that an understanding of education's context, policy, and practice will, at best, be a narrow one, and, perhaps, at worst, a flawed one. We would argue therefore that making *space* for space in education is long overdue.

NOTES

1. Comparative education has certainly investigated issues of geography, but with a different scale of significance, i.e. comparisons between national education systems.
2. After earlier translations into Italian (1975) and Japanese (1975).
3. For example, see two recent special issues of journals devoted to the role of post-structuralism in education policy studies (e.g. *Journal of Education Policy*, 18.2, 2003), and geography *(Environment and Planning A*, 36, 2004).
4. Some of this work on topics concerning space and education has been a part of symposia at various international conferences (e.g. American Educational Research Association, British Educational Research Association), included in journal papers, or as special editions of journals (e.g. *Forum: For Promoting 3–19 Comprehensive Education*, 46.1, 2004).

REFERENCES

Agnew, J. (2005) 'Space: place', in P. Cloke and R. Johnston (eds.), *Spaces of geographical thought: deconstructing human geography's binaries*, London: Sage.

Appadurai, A. (1996) *Modernity at large: cultural dimensions of globalization*, Minneapolis: University of Minnesota Press.

Armstrong, F. (2003) *Spaced out: policy, difference and the challenge of inclusive education*, London: Kluwer.

Ball, S. J. (1998) 'Big policies/ small world: an introduction to international perspectives in education policy', *Comparative Education*, 34.2: 119–130.

—— (2003) 'The risks of social reproduction: the middle class and education markets', *London Review of Education*, 1.3: 163–175.

—— (2004) 'The sociology of education: a disputational account', in S. J. Ball (ed.), *The RoutledgeFalmer reader in sociology of education*, London: RoutledgeFalmer.

—— (2006) 'The necessity and violence of theory', *Discourse: studies in the cultural politics of education*, 27.1: 3–10.

Ball, S. J., Bowe, R. and Gewirtz, S. (1995) 'Circuits of schooling: a sociological exploration of parental choice in social class contexts', *Sociological Review*, 43: 52–78.

Ball, S. J., Maguire, M. and Macrae, S. (1998) '"Race", space and the further education market place', *Race, Ethnicity and Education*, 1.2: 171–189.

Barnes, T. J. and Hannah, M. (2001) 'The place of numbers: histories, geographies and theories of quantification', *Environment and Planning D*, 19: 379–383.

Bauman, Z. (2000) *Liquid modernity*, Cambridge: Polity Press.

Benito, A. E. (2003) 'The school in the city: school architecture as discourse and as text', *Pedagogica Historica*, 39.1–2: 53–64.

Benjamin, W. (1999) *The Arcades project*, Cambridge, MA: Belnap Press.

Bourdieu, P. (1999) 'Site effects', in P. Bourdieu, A. Accardo, G. Balazs and P. P. Ferguson (eds.), *The weight of the world: social suffering in contemporary society*, Cambridge: Polity Press.

Callejo Pérez, D. M., Fain, S. M. and Slater, J. J. (eds.) (2004) *Pedagogy of place: seeing space as cultural education*, New York: Peter Lang.

Casey, E. S. (1996) 'How to get from space to place in a fairly short stretch of time: phenomenological prolegomena', in S. Feld and K. H. Basso (eds.), *Senses of place*, Sante Fe, NM: School of American Research Press.

Cox, K. R. (1998) 'Spaces of dependence, spaces of engagement and the politics of scale or: looking for local politics', *Political Geography*, 17.1: 1–23.
Crang, M. and Thrift, N. (2000) 'Introduction', in M. Crang and N. Thrift (eds.), *Thinking space*, London: Routledge.
Dale, R. (1999) 'Specifying globalization effects on national policy: a focus on mechanisms', *Journal of Education Policy*, 14.1: 1–17.
Deleuze, G. and Guattari, F. (1987) *A thousand plateaus: capitalism and schizophrenia*, Minneapolis: University of Minneapolis Press.
Eagleton, T. (1997, December 5) 'International books of the year', *Times Literary Supplement*, p.11.
Edwards, R. and Usher, R. (eds.) (2003) *Space, curriculum and learning*, Greenwich, CT: Information Age Publishing.
Ellsworth, E. (2005) *Places of learning: media, architecture, pedagogy*, London: Routledge.
Falconer Al-Hindi, K. (2001) 'Do you get it? Feminism and quantitative geography', *Environment and Planning D*, 19: 505–513.
Foucault, M. (1977) *Discipline and punish: the birth of the prison*, London: Penguin.
Gregory, D. (1994) *Geographical imaginations*, Oxford: Blackwell.
Gulson, K. N. (2005) 'Renovating educational identities: policy, space and urban renewal' *Journal of Education Policy*, 20.2: 147–164.
——— (2006) 'A white veneer: education policy, space and "race" in the inner city', *Discourse: studies in the cultural politics of education*, 27.2: 251–266.
Hall, E. (1999) 'Workspaces: refiguring the disability-employment debate', in R. Butler and H. Parr (eds.), *Mind and body spaces: geographies of illness, impairment and disability*, London: Routledge.
Haraway, D. J. (1999) 'Situated knowledges: the science question in feminism and the privilege of partial experience', in B. Mario (ed.), *The science studies reader*, New York: Routledge.
Harvey, D. (1993) 'From space to place and back again: reflections on the condition of post-modernity', in J. Bird, B. Curtis, T. Putnam, G. Robertson and L. Tickner (eds.), *Mapping the futures: local cultures, global change*, London: Routledge.
Henry, M., Lingard, B., Rizvi, F. and Taylor, S. (2001) *The OECD, globalisation and education policy*, Oxford: Permagon.
Hubbard, P., Kitchin, R. and Valentine, G. (2004) 'Editor's Introduction', in P. Hubbard, R. Kitchin and G. Valentine (eds.), *Key thinkers on space and place*, London: Sage.
Kincheloe, J. L. (2001) 'Describing the bricolage: conceptualising a new rigor in qualitative research', *Qualitative Inquiry*, 7.6: 679–692.
King, A. (ed.) (1980) *Buildings and society: essays on the social development of the built environment*, London: Routledge and Kegan Paul.
Kuhn, T. S. (1962) *The structure of scientific revolutions*, Chicago: University of Chicago Press.
Lawn, M. and Grosvenor, I. (eds.) (2005) *Materialities of schooling: design-technology-objects-routines*, Oxford: Symposium.
Leander, K. M. and Sheehy, M. (eds.) (2004) *Spatializing literacy research and practice*, New York: Peter Lang.
Lefebvre, H. (1991) *The production of space*, Oxford: Blackwell.
Levin, B. (1998) 'An epidemic of education policy: (what) can we learn from each other?' *Comparative Education*, 34.2: 131–141.
Markus, T. (1993) *Buildings and power: freedom and control in the origin of modern building types*, London: Routledge.
Marsden, W. E. (1977) 'Historical geography and the geography of education', *History of Education*, 6.1: 21–42.

——— (1987) *Unequal educational provision in England and Wales: the nineteenth century*, London: Woburn Press.
Massey, D. (1992) 'Politics and space/time', *New Left Review*, 196: 65–84.
——— (1993a) 'Politics and space/time', in M. Keith and S. Pile (eds.), *Place and the politics of identity*, London: Routledge.
——— (1993b) 'Power-geometry and a progressive sense of place', in J. Bird, B. Curtis, T. Putnam, G. Robertson and L. Tickner (eds.), *Mapping the futures: local cultures, global change*, London: Routledge.
——— (1994) *Space, place and gender*, Cambridge: Polity.
——— (2005) *For space*, London: SAGE Publications.
McGregor, J. (2004) 'Space, power and the classroom', *Forum: for promoting 3–19 comprehensive education*, 46.1: 13–18.
Mourad, R. P.(1997) 'Post-modern interdisciplinarity', *The Review of Higher Education*, 20.2: 112–140.
Saint, A. (1987) *Towards a social architecture: the role of school building in post-war England*, New Haven, CT.: Yale University Press.
Seaborne, M. (1971) *The English school: its architecture and organisation 1370–1870*, London: Routledge and Kegan Paul.
Seaborne, M. and Lowe, R. (1977) *The English school: its architecture and organisation, Vol. 2, 1870–1970*, London: Routledge and Kegan Paul.
Shields, R. (1991) *Places on the margin: alternative geographies of modernity*, London: Routledge.
Smith, N. (2004) 'Space and substance in geography', in P. Cloke, P. Crang and M. Goodwin (eds.), *Envisioning human geographies*, London: Arnold.
Smith, N. and Katz, C. (1993) 'Grounding metaphor: towards a spatialised politics', in M. Keith and S. Pile (eds.), *Place and the politics of identity*, London: Routledge.
Soja, E. W. (1996) *Thirdspace: journeys to Los Angeles and other real-and-imagined places*, Cambridge, MA: Blackwell.
Symes, C. (1996) 'Building the Queenslander: the contribution of school architecture to the formation of the child', in L. Finch (ed.), *Young in a warm climate: issues in Australian childhood*, St. Lucia, Australia: University of Queensland Press.
——— (1999) 'First impressions: semiotics of school vestibules', in C. Symes and D. Meadmore (eds.), *The extra-ordinary school: parergonality and pedagogy*, New York: Peter Lang.
Taylor, C. (2001) 'Hierarchies and "local" markets: the geography of the "lived" market place in secondary education provision', *Journal of Education Policy*, 16.3: 197–214.
Thrift, N. and Whatmore, S. (2004) 'Introduction', in N. Thrift and S. Whatmore (eds.), *Cultural geography: critical concepts in the social sciences Vol 1*, London: Routledge.
Urry, J. (2000) 'Mobile sociology', *British Journal of Sociology*, 51.1: 185–203.
——— (2005) 'Mobility and proximity', *Sociology*, 36.2: 255–274.
Vincent, C., Ball, S. J. and Kemp, S. (2004) 'The social geography of childcare: making up the middle-class child', *British Journal of Sociology of Education*, 25.2: 229–244.
Vincent, C. and Martin, J.(2002) 'Class, culture and agency', *Discourse: studies in the cultural politics of education*, 23.1: 109–128.
Zukin, S. (1998) 'Urban lifestyles: diversity and standardisation in spaces of consumption', *Urban Studies*, 35.5–6: 825–839.

2 The spatial politics of educational privatization

Re-reading the US homeschooling movement

Claudia Hanson Thiem

Over the course of two decades, homeschooling in the United States has evolved from an isolated and underground practice into an organized, visible, and confident movement. Once expressly forbidden in many parts of the country, homeschooling was legalized in all fifty states by 1993. Ten years later, according to U.S. Department of Education estimates, approximately 1.1 million children were receiving at least a portion of their education at home (NCES 2004). Advocates at the National Home Education Research Institute claim numbers up to twice as large, but even the more conservative figure constitutes a remarkable 2.2 percent of the nation's school-aged population. These numerical gains, while impressive, do not fully capture the significance of the homeschooling phenomenon. In addition to growing a base of dedicated, often fervent practitioners, the homeschooling movement has developed a substantial organizational infrastructure. The emergence of professional advocacy groups, specialized publishers, and even institutions of higher education, to say nothing of thousands of informal local support groups, signals the institutionalization of homeschooling in the landscape of US educational choice (Stevens 2001). Finally, the movement's 'coming of age' is evident in its growing acceptance among non-practitioners. While a majority of Americans surveyed in 2001 still disapproved of homeschooling, 41 percent considered it 'good for the nation'—up from only 16 percent in 1985 (Gallup 1985; Rose and Gallup 2001). Congressional declarations of 'National Home Education Week' in both 1999 and 2000 cast a positive light on the practice, while growing attention from major media have nudged it towards the cultural mainstream. As if to confirm homeschooling's arrival, critics now describe it as a serious threat to common schooling—alongside voucher programs, curricular opt-outs and other forms of educational privatization (Apple 2001; Bivins 2003; Lubienski 2000; Reich 2002; Riegel 2001).

How can we explain the rapid rise and institutionalization of home education? The academic literature around this question is sparse, but the story that emerges generally features the interplay between a crisis-ridden public education system and a robust evangelical Christian movement. Survey research provides some indirect support for this narrative, with parents listing educational quality and religious convictions as their primary reasons for homeschooling (Bielick et al. 2001; Mayberry 1988). Such individual preferences, however, do not automatically produce a 'successful' movement. Thus, to account for homeschooling gains, particularly the relaxing of legal restrictions around the practice, political and sociological research has examined the movement's organizational strategies and ideological affinities. Stevens, for example, describes how homeschoolers (particularly evangelical Christians) have built disciplined advocacy networks that can be mobilized for political action (Stevens 2001; Cibulka 1991). Apple, meanwhile, argues that homeschooling has benefited from its resonance with broader pro-market and anti-statist educational politics, which themselves have gained influence in the United States in recent years (Apple 2001).

While the above are certainly foundational elements of any account of homeschooling's development, I argue that deploying a spatial perspective on political strategies can enrich existing narratives. In particular, exploring the ways in which homeschoolers have constructed place, space, and scale, and how these constructions have articulated with the geographies of educational regulation, reveals additional dimensions of the movement's growth. This claim is a familiar one within human geography, where scholars of social movements and other forms of organized 'resistance' rehearsed the basic contours of a spatialized political analysis in the 1990s. Building on constructivist ontologies of space, these critical researchers rejected the treatment of geography as a mere backdrop to, or container for political mobilizations. Instead, they proposed a reanimated geography of politics, in which space, place, and scale are both the object and medium of political struggle (e.g. Delaney and Leitner 1997; Massey 1994; N. Smith 1993; Soja 1989; Staeheli 1994; Swyngedouw 1997). Political struggle still unfolds in place and across space, but these geographies—for example, the 'contents' of places and scales, and their relations across space—can themselves be manipulated to facilitate or constrain particular outcomes. The redefinition (both material and discursive) of space, place, and scale is a powerful political strategy that merits consideration in educational research.

This chapter, then, develops the homeschooling case as a basic template for a spatialized political analysis. To this end, I explore two forms of spatial political practice, centred on place and scale, respectively, that appear central to the homeschooling movement's successes.[1] First, I examine how practitioners have strategically redefined the private home as a protected, and indeed *privileged* site for educating children. In doing so, they tap into a longstanding discourse of private property rights and the more recent valorization of localized or family-based social provision under neoliberalism. Second, I consider the gains realized from networking dispersed homeschooling spaces. The development of national political, cultural, and economic infrastructures has provided strategic advantages for working in a spatially fragmented (federalist) regulatory environment. But before elaborating these examples, I provide a brief conceptual introduction to the politics of place and scale. Rather than engaging deeply with the work of particular geographical theorists, this sketch develops a broad foundation for a constructivist spatio-political imaginary, that is, a methodological sensitivity that might inform a wide range of empirical inquiry.

SPATIALIZING POLITICS

The political geography of homeschooling developed below starts from the assertion of critical human geographers—now a truism within the discipline—that space, place and scale are social constructs. This claim originated in the spatialized Marxisms of Harvey (1982, 1985), Lefebvre (1991) and Soja (1980), who variously related the organization of space to processes of capital accumulation and class struggle. One oft-cited concept, Soja's 'socio-spatial dialectic', effectively conveys the emerging constructivist sensibility. 'The structure of organized space', he proposed, 'is not a separate structure with its own autonomous laws of construction and transformation…it represents, instead, a dialectically defined component of the general relations of production, relations which are simultaneously social and spatial' (Soja 1980: 208). Smith, meanwhile applied a similar materialist analysis to geographical scale, arguing in particular that the national scale represents a historically-specific compromise between the imperatives of inter-capitalist competition and cooperation (N. Smith 1984). Subsequent work positioned additional scales such as the body, the urban, and the global, in relation to processes of capital accumulation (N. Smith 1993; Taylor 1982). While these Marxian formulations effectively challenged the treatment of geography as a fixed grid or container, they were subsequently criticized for a narrow reading of the 'production' of space and scale. In a movement to de-centre capital and class as privileged forces, later interventions added race and gender relations, cultural practices, language, and performance to the spatial constructivist imagination (Keith and Pile 1993; Marston 2000; Massey 1994; Thrift 1996, 1999).

While such disputes about the assignment of agency are ultimately irreconcilable, what unites spatial constructivist ontologies, and in turn underpins a spatialized politics, is the recognition that geographies are both contingent and malleable. Put simply, if geographies are socially produced and productive of social relations, there exist both opportunities and incentives to manipulate and contest them. To clarify this rather abstract notion, the remainder of this section elaborates common treatments of place and scale that appear in geographical analyses of political mobilizations.

What is the relationship between place and politics? While theorizations of this couplet are too numerous to review here in detail, Staeheli (2003) identifies two broad approaches that are particularly useful for framing the homeschooling story. First, place, which may be defined for the moment as a physical location, may become the object of turf battles. Control of place may provide access to political resources, aid in mobilizations, or even constitute a precondition for 'everyday' social practices. As one example of such a politics 'about' place, Staeheli points to anti-development campaigns by residential property owners. So-called NIMBY movements see territorial control as a means of preserving economic assets. Control of place may also support 'externally' oriented political projects. Thus, Routledge's mapping of Nepal's pro-democracy revolution describes how activists worked in spaces that were inaccessible to the repressive and surveillance organs of the state (Routledge 1997 cited in Staeheli 2003). Beyond geography, the work of bell hooks might also be read in this vein. In her essay 'Homeplace: a site of resistance', hooks describes how the black home has allowed for the creating of caring relationships, themselves a powerful political asset, in an oppressive world (hooks 1990).

The politics of place, however, involve more than efforts to control critical pieces of territory. If we take seriously Massey's relational re-reading of place, the geographical 'stakes' in such struggles are not pre-given entities. Rather, they are themselves the dynamic (and contested) articulations of social relations in space-time (Massey 1994, 2005). Accordingly, a second common approach to the place–politics nexus considers the (re)construction of place itself—its contents, boundaries, and constitutive relations—as a form of political action (Staeheli 2003). To illustrate this, Staeheli works through a feminist geographical critique of the public sphere. The historic exclusion of women from politics, she argues, cannot be separated from the gendered construction of public space. Indeed, 'it is often the characteristics of *places*—the physical and social characteristics and meanings—that deny or limit access to certain types of people and certain types of behaviour at certain times and thereby limit or constrain the "public"' (Staeheli 2003: 166, emphasis added). Thus, remaking public spaces to admit women has been central to an effective feminist politics. In this context, modifying the cultural norms that regulate public spaces has resulted in important gains for women. But winning material changes such as adding lavatories and lighting have also increased access. Beyond traditional public spaces, parallel efforts to use place to empower women can be seen in efforts to redefine the home as a site of political agency and open up its contents to public scrutiny (e.g. Staeheli 1996; Brown 1997; Marston 2000).

Geographical scale has similarly been analyzed as a political tool. Here, one of the first and most influential discussions can be found in Smith's work on 'jumping scale' (N. Smith 1992; 1993). This term, introduced in the context of progressive mobilizations, described the pursuit of political action in spaces 'beyond' those where particular grievances or bases of power were anchored. Smith stresses a local-to-global movement, where ostensibly localized political struggles are moved to higher levels of the state, or pursued over larger (areal) spaces.

This initial formulation has been extended and reworked in several directions. To begin, subsequent discussions have emphasized that scale-strategies are neither unidirectional nor the exclusive domain of progressive actors. Studies of neoliberalization, for example, identify a strategic 'localization' of politics in instances of regulatory decentralization (Brenner 2004; Peck 2002; for an excellent education-based case study, see Robertson and Dale 2002). The state itself is thus an active practitioner of a scalar politics, and its constructed geographies structure (but do not fully contain) 'external' mobilizations. Other researchers have proposed network ontologies of scale to complement or challenge Smith's hierarchical model. Cox (1998), for example, conceptualizes scale politics as the negotiation of separate 'spaces of dependence' and 'spaces of engagement'. The former are spaces on which groups 'depend for the realization of essential interests and for which there are no substitutes elsewhere' (p. 2), while the latter are the centres of social power (often state agencies) where the spaces of dependence must be secured. The politics of scale thus involve constructing 'networks of association' between place-dependent actors and allies who may be better positioned in relation to spaces of engagement (e.g. Cox 1998; Leitner et al. 2002). Finally, some theorists have explored scale's epistemological dimensions. Scale, they note, provides a convenient short-hand for narrating events; we commonly think in terms of 'local' politics, 'national' interests, 'global' flows, and so on. Critically, such scale framings, although they carry considerable political power, have no *inherent* connections to the material spatialities of the processes they describe. Representations of scale, particularly as they invoke appropriate venues and audiences for issues, are thus also both stake and strategy in political struggle (Delaney and Leitner 1997; Jones 1998; Swyngedouw 1997).

Two case studies of social movement mobilization demonstrate some of scale's varied analytic possibilities.[2] In Miller's analysis of Boston-area anti-nuclear activism, part of a larger effort to spatialize social movement theory, local peace organizations faced a common scale dilemma: while their grievances derived from national policies, the local state (Massachusetts) afforded a more open political structure (Miller 1994, 1997, 2000). This was not, however, a purely constraining factor. Indeed, activists were able to portray anti-militarism as a local issue, and won the passage of 'nuclear-free zone' resolutions in several municipalities. But these victories remained largely symbolic; a proposed *national* nuclear freeze foundered in Congress because of its threats to regional defence economies. Miller's analysis, then, demonstrates how the scaling of both political discourse and engagement affect movement outcomes. He further positions the spatialities of political opportunity structures (i.e. Massachusetts in this instance) as a key consideration for a scalar politics.

Bosco, meanwhile, folds scale into a discussion of the network spatiality of collective action (Bosco 2001). His work explores the successes of Argentina's Madres de Plaza de Mayo, a human rights group formed in

1977 to protest the state's kidnapping and torture of dissidents. The Madres growth and longevity, Bosco argues, can be attributed to the operation of 'geographically flexible networks' that have, over time, addressed a variety of spatial constraints. For example, Madres groups throughout Argentina participate in weekly ritual walks around their local town squares—a symbolic reproduction of the group's original protests in the Plaza de Mayo in Buenos Aires. This shared practice combats the physical distance between members, and also creates local points of reference for dispersed supporters. A second key aspect of network strategy involved seeking support abroad. During their early years, the Madres faced particularly repressive conditions at home. But relationships with expatriated Argentinians in Europe provided a bridge to transnational human rights networks, and eventual access to legal, financial, and public relations support. These instances of spatial networking are similar to scale-jumping—but less hierarchical in conceptualization or practice—and represent strategic responses to the changing (and geographically uneven) contexts for political mobilization.

Shades of these spatial strategies and practices will reappear in the discussion of homeschooling that follows, although no particular formulation provides an adequate 'map' of the movement's development. Thus, in introducing the politics of place and scale, my intent has simply been to suggest the lines of analysis a constructivist geographical imagination might support. Moving forward into the case study, recalling the broad malleable nature of spatial organization will be more important than the specificities of individual theoretical constructs.

A POLITICAL GEOGRAPHY OF HOMESCHOOLING

Before considering *how* homeschooling has become a significant feature on the educational landscape, it may be useful to review the movement's life-course and vital statistics. Modern home education originated in the counter-cultural movements of the late 1960s and early 1970s, when pioneering parents rejected the bureaucratic rigidities and authoritarian practices of institutional education.[3] Drawing inspiration from progressive educational philosophies and the nascent alternative schools movement, early homeschoolers experimented with child-centred pedagogies and the more radical idea of 'unschooling', i.e. the separation of education from formal instruction. But if contemporary homeschooling originated on the political and cultural left, it soon attracted a right-leaning, religious contingent. Troubled by the growing secularization of public education and diminished access to Christian schools, increasing numbers of conservative Protestants adopted homeschooling to control peer influences and reintegrate religious training into their children's lives.[4] For many of these ideologically-motivated practitioners, the decision to homeschool also reflected their belief in the (nuclear) family as a privileged social institution (Carper 2000; Van Galen 1991).

Homeschooling, in its early years, was a quiet affair. In 1980, compulsory attendance laws explicitly forbade the practice in thirty US states, and parents risked criminal charges of truancy or neglect if discovered by state authorities. As a result, many homeschoolers hid their activities from public view; they shuttered windows during normal school hours, kept children indoors, and rehearsed cover stories for inquisitive friends and neighbours. While this secrecy obviously hindered attempts to count practitioners, researcher-advocate Patricia Lines used indirect methods to estimate fifteen thousand active homeschoolers in the early 1970s (Lines 1998). The subsequent growth of this number paralleled the gradual, state-by-state legalization of homeschooling (see Table 2.1), a process which involved both litigation and legislative action. By 1993, parents could legally educate their children at home (subject to regulation) in all fifty US states. The number of homeschooled students, meanwhile, climbed into the hundreds of thousands. Homeschooling in its (mature) phase has been marked by two notable trends. First, the movement has built an impressive organizational infrastructure, which has both facilitated further growth and solidified political gains. Second, homeschooling has witnessed a gradual (and highly contested) rapprochement with the state. Increasing numbers of public schools offer partnership programmes, and parents can access a variety of state aids such as tax credits, special education services, and in some cases, charter school funds.

Throughout this history, the early divisions between progressive reformers and religious dissenters—so-called 'pedagogues' and 'ideologues', or 'inclusives' and 'believers'—have remained (Stevens 2001; Van Galen 1991). Based on their motives for homeschooling, parents join specialized support groups, subscribe to different publications, attend separate conferences, and often spar over the legitimate representation of their interests. Thus, it is in many ways misleading to speak of a unified movement. While reliable counts of homeschooling's various communities are unavailable, conservative Protestants and the organizations that support them, often speak with

Table 2.1 Numbers of homeschooled students by year

Year	*Homeschooled students*
1980	10,000
1985	50,000
1990	300,000
1995	700,000
1999	850,000
2003	1,100,000

Sources: Bielick et al. (2001), Lines (2000), NCES (2004)

the loudest voice in homeschooling politics (Carper and Ray 2002). Their activities, as a result, feature prominently in the analysis below.

Now, constructing a geographical account of homeschooling requires conceptualizing the movement's development as, at least in part, a spatial problematic. This is not a difficult exercise, and indeed, three such challenges appear in the sketch above. Establishing the home as a socially and legally acceptable site for providing and governing education, mobilizing dispersed practitioners, and negotiating the fragmented and uneven regulatory terrain of US federalism have all been important moments in homeschooling's institutionalization. These problems, then, constitute the 'targets' of the spatial strategies described below.

'HOME' AND 'SCHOOL' IN HOMESCHOOLING

The practice of homeschooling hybridizes two 'distinct' sites of social reproduction: home and school. While many advocates promote this mix as a natural one, the thousands of 'homeschools' in the United States are precarious spatial constructs, potentially unsettling to both the ideal-typical places from which they are formed. One key challenge involves legitimating the homeschool. To gain acceptance, practitioners must justify relocating education from schools, which are state-sanctioned spaces for instilling academic and social skills. Once established, the home-as-school may invite scrutiny by state authorities. Because familial sovereignty underpins many decisions to homeschool, this potential violation of the home's traditional social and physical boundaries is particularly troubling (e.g. Allan and Crow 1989; Saunders and Williams 1988). Control of the home, including its relations with various 'external' spaces, is a crucial prerequisite for homeschooling freedoms. Homeschooling advocates have thus practised a politics of place in their efforts to define and defend the homeschool construct. Indeed, strategic reconstructions of the home, both material and discursive, have often complemented homeschoolers more familiar demands for parental rights.

A common approach, a ubiquitous phrase in homeschoolers' rhetoric, involves contrasting traditional schools with their home-based counterparts. Both ideologically- and pedagogically-motivated practitioners, for example, describe the home as a nurturing and stimulating place for learning. It is, by their accounts, more conducive to children's development than traditional classrooms. While parents' ability to control the homeschool setting obviously facilitates this advantage, homeschoolers are quick to contest depictions of the home as an isolated place. To counter concerns about children's socialization, they boast of frequent fieldtrips, social gatherings, and club memberships, and regular work in the community (see C. Smith and Sikkink 1999). Some homeschoolers supplement these comparative rhetorics by reproducing familiar elements of the school.

During a designated 'school day,' certain areas of the home *become* traditional classrooms; spaces may be outfitted with desks (for both students and teacher), chalkboards, and perhaps, the flag. Such replications are less common among liberal homeschoolers, and may represent more of an ideal than actual practice even among conservatives (Stevens 2001). Even so, the use of familiar tropes to describe the homeschool may enhance its legitimacy as an educational space.

Other arguments for relocating education eschew institutionalized schooling rather than selectively appropriating its goals and practices. The social relations that constitute traditional classroom settings damage children, some advocates claim. While part of the original case for unschooling and other liberal pedagogical practices, this indictment of schools is most systematically elaborated and deployed in conservative Protestant branches of the movement. Citing Biblical authority, many Christian homeschoolers view the home as an inviolable nexus of God and the traditional family. As a sacred responsibility expressly assigned to parents (and ultimately, fathers), education thus *properly* belongs in the home. Schools, by contrast, represent a corrupt external world. Public schools, in particular, espouse secular humanist values and expose children to dangerous peer-group influences. Alarmist books such as *Lambs to the Slaughter*, *Child Abuse in the Classroom*, and *The Harsh Truth about Public Schools* target complacent parents, and similar warnings circulate in a multitude of online discussion forums. These particular constructions, aptly described by Apple (2001) as 'Satan's threat versus the fortress home', are more likely to deepen existing support than convince homeschoolings' detractors. Nevertheless it reveals another dimension of the discursive reconstruction of place underway within the movement.

Even if the home gains acceptance as an educational space, surveillance and regulation can limit homeschoolers' curricular and pedagogical choice. Thus, conservative homeschoolers have done more than deploy a spatial separatist language, but have further attempted to selectively harden the home's material boundaries with the 'outside' world. Many advocates adopt a hyper-vigilant stance against perceived intrusions. One issue of *The Teaching Home* (a popular Christian homeschooling magazine), for example, reminds fathers of their responsibility to 'promote public policies which respect the sanctity of the home' (cited in Diamond 1997: 181). Meanwhile, the Home School Legal Defense Association (HSLDA) has argued consistently for strict limits on access to the home, either by education officials or social services personnel (HSLDA 1998). Violations of Fourth Amendment rights (against unwarranted search and seizure) and other apparent threats to privacy are a recurrent theme in the group's publications and advisories, and are presented as an emerging legal battleground.

Of course, these strategic reconstructions of home and school do not automatically establish legitimacy or rights for homeschoolers. Instead, their effectiveness depends on how they articulate with state strategies and spatial capabilities, and the evolving *legal* construction of the home. To the movement's advantage, claims to the private home resonate with the ideals of individual freedom, privacy, and property rights—all foundational values of American liberalism.[5] Housing scholars Saunders and Williams describe the political-cultural construction of the British home in a passage that translates easily to the American context:

> the idea of the home [is] as the private sphere. In the polarity between state and society, public and private, collectivism and individualism, the home stands as the symbolic antithesis to the state. It is around the idea of the home that the ideals of freedom, autonomy and self-determination are reproduced. The State may be a fortress, but the home is, as the saying goes, a castle. (Saunders and Williams 1988: 88; see also Rykwert 1991)

The home, then, has historically been treated as a legally private space, a sort of regulatory no-go zone for the state.

Despite interrogations and strategic openings by feminist critics, the idealized private home may actually be regaining salience and 'closure' in the developing neoliberal policy environment. Long an ideological buffer against socialized welfare provision (e.g. Munro and Madigan 1999), the family-home complex is revalorized in the context of state retrenchment and individualized responsibility (e.g. Bakker 2003; Brown 2004). In education, the home is central to recent narratives of crisis (e.g. 'broken homes contribute to low achievement') and life-long learning (e.g. formal schooling is insufficient preparation for work in a knowledge economy). Homeschooling further resonates with other forms of spatial separatism that pervade contemporary cultural politics: 'It is an extension of the "suburbanization" of everyday life...the equivalent of gated communities and of the privatization of neighbourhoods, recreation, parks, and so many other things' (Apple 2001: 175). In this supportive environment for privatized social reproduction, the particularly 'negative' tone of homeschoolers' demands may also help their cause. Most practitioners desire to be 'left alone' by authorities, rather than requesting state support (beyond basic recognition) (see Cibulka 1991). Regulating homeschools, in any case, presents a practical challenge to increasingly lean state agencies. Their dispersed locations, irregular schedules, and heterogeneous functions make rigorous monitoring difficult.

While homeschoolers have managed the homeschool hybrid to their advantage thus far, their control and definitions of this place are not fixed or immutable. Indeed, the increasing popularity of 'virtual charter schools' and related state-accredited home-study programs highlights the vulnerability of established boundaries. The virtual charter model offers state-funded distance education to would-be homeschoolers; curriculum materials, professional teacher consultants, and even computing equipment are provided at no cost to enrolled students (see Huerta and Gonzalez 2004). Although most homeschooling still occurs independently, critics within the movement worry that the growth of virtual programs—and their promotional use of the term 'homeschooling'—invites regulation of 'legitimate' homeschooling practice. Such schools literally bring the state into the home via textbooks, assessments, and regular telephone conferences. Opponents argue that these intrusions endanger more than just participating households. By normalizing, however slightly, a porous and 'surveillable' home, they threaten the foundations of the movement at large (see Klicka 2002; We Stand for Homeschooling n.d.).

HOMESCHOOLING BEYOND THE HOME

While redefining home has been an important strategy for homeschoolers, the movement has also gained from networking beyond these 'spaces of dependence' (Cox 1998).[6] Scale strategies and network formations have helped practitioners and advocates negotiate the political environments in which they operate—and, in particular, manage the complexities of decentralized educational regulation. Mapping the organizational infrastructure of homeschooling reveals two particular forms of spatial practice that have advanced the movement in this regard. First, homeschoolers have formed state and national advocacy groups, whose specialized activities reflect a highly nuanced reading of the fragmented regulatory terrain. In addition, a burgeoning cultural-economic complex has assisted in the networking of homeschooling space. Although less overtly political than the advocacy infrastructure, the common fields of reference constructed by media, vendors, and related enterprises strengthen community and preserve channels for potential mobilizations. I sketch these moments of 'scale-jumping' and spatial networking in turn.

Most battles over the legal status of homeschooling have occurred in the court systems and legislatures of individual US states, which retain responsibility for educational regulation within the federal system. Not surprisingly, the current organizational infrastructure of homeschooling mirrors this scalar division of governance: at least one Christian and one 'inclusive' statewide coordinating group can be found in most states. This congruence, however, did not develop automatically. While networking is a longstanding feature of modern homeschooling, early practitioners generally formed *local* support groups, whose concerns were more often pedagogical than political. Many state-level organizations formed initially to address a regulatory void: or, more precisely, an environment that produced irregular and unpredictable contacts with local education officials. In the language of the politics of scale, these groups sought to strategically *produce* a state-level regulatory structure or 'space of engagement' where none existed previously (Bates 1991; Cibulka 1991; Stevens 2001; Tyler 2003;

on scale see Cox 1998; Herod 1991). Their activities were instrumental in establishing or codifying homeschooling rights in the 1980s. Since legalization, state-level groups have continued to play an essential role in organizing the movement. While they undertake important community-building functions such as sponsoring events and connecting new homeschoolers to support groups, their particular contribution to a spatial politics derives from their rich stores of 'local' knowledge. State homeschooling organizations are steeped in the legal particularities and legislative cultures that structure regulation (see Table 2.2), a quality which enhances their value to members, and makes them powerful political advocates.

Given the fragmented geographies of US educational regulation, homeschooling may have easily evolved into a loose aggregate of statewide movements, rather than the ostensibly national cultural-political force that exists today. Homeschoolers' activities 'above' or 'beyond' the states should not be confused with a desire for a national regulatory structure. Indeed, most feel this would do more to threaten than advance their rights. Nevertheless, organizing nationally presents several advantages in negotiating policy environments and political opportunity structures that vary widely over space (and time). Limited federal authority over education, meanwhile, assures homeschoolers that the federal government will not similarly 'scale up' its activities. In this context, national advocacy groups, particularly those associated with conservative Protestant homeschooling, have aided practitioners' quest for legality and legitimacy.[7] Two in particular merit extended discussion despite their controversial positions within the movement as a whole. The Home School Legal Defense Association (HSLDA) and National Home Education Research Institute (NHERI) have respectively organized and represented homeschoolers in ways that effectively respond to, and even take advantage of the varied state-level regulation of home education.

The HSLDA was established in 1983 to provide a form of pooled legal insurance for homeschoolers. For a modest annual fee, it guaranteed members counsel in unwelcome encounters with state authorities. Never a purely reactive organization, the HSLDA participated in many state-level efforts to legalize homeschooling. In such struggles, its 'national' character provided a dual benefit: lawyers had the flexibility to select and pursue the most promising cases, while the movement as a whole could be advanced strategically on multiple fronts. In addition to providing legal services to advance the movement, the HSLDA has cultivated an attentive and disciplined grassroots political network. In 1990, it established the National Center for Home Education (NCHE), a special division devoted to political organizing. NCHE's 'legislative toolbox' allows web visitors to track developments in their home states and provides contact information for relevant elected officials. Its Congressional Action Program trains lead contacts in all 435 Congressional districts to coordinate responses to federal legislation. The NCHE has also developed a selective network of state affiliates

Table 2.2 Selected homeschooling regulations by state

State	*Level of regulation*	*Specific requirements*
Illinois	None	Parents are not required to notify state authorities of their decision to homeschool. Core subjects should be taught, but homeschools do not submit student work or test scores for review. There are no attendance requirements or minimum educational qualifications for parent-instructors.
California	Low	Independent homeschools are regulated as private schools. Parents must notify state-level authorities annually of their intent to homeschool. English-language instruction in typical public school subjects must be provided, but no records of attendance, student work, or testing are required. Parents must be 'capable of teaching'.
Georgia	Moderate	Parents must file a declaration of intent to homeschool with local authorities. Homeschools must be in session 180 days per year for 4.5 hours daily, and provide instruction in core subjects. Attendance records, monthly reports, annual statements of progress, and standardized test scores (every third year) must be submitted for review. Parent-instructors must hold a high school diploma or GED equivalent.
North Dakota	High	Parents must provide annual notification of homeschool operation to local authorities. Homeschools must be in session 175 days per year for 4 hours daily. An extensive list of subjects—parallel to public school curricula—must be taught. Parents must submit records of courses taught, results of periodic assessments, and bi-annual standardized test scores. Students scoring beneath the 50th percentile are targeted for remediation. Parent-instructors must hold a teaching certificate or 4-year college degree. Those with a high school diploma or equivalent may begin homeschooling under the supervision of a certified teacher.

Source: Home School Legal Defense Association, 2006 (http://www.hslda.org/laws)

that collect information locally and help disseminate alerts to supporters. While none of these are particularly innovative organizing techniques, they perform a vital function for homeschooling politics. The NCHE's monitoring and mobilizing capabilities allow movement leadership to target resources—legal, financial, or human—to particularly urgent struggles.

These partner organizations demonstrated their ability to marshal homeschoolers nationally in a memorable 1994 battle over proposed federal teacher certification requirements. Concerned that pending legislation might be applied to homeschoolers, the HSLDA and NCHE sent urgent bulletins to their state-level affiliates and staged a media blitz. In the following weeks, home education supporters placed hundreds of thousands of calls to Congressional offices, several times overloading telephone switchboards at the Capitol. Beyond producing a favourable amendment, this show of strength established the movement as a formidable political force. The work of organizing, of course, continued beyond this victory and the HSLDA/NCHE remains attentive to the local dependencies of their eighty thousand member-families. The organizations' pronouncements, though hyperbolic, suggest an implicit recognition of the importance of spatial flexibility (Bosco 2001). As their president J. Michael Smith warned in a *Washington Times* editorial: '...the stakes are high for homeschoolers. One victory for opponents of homeschooling in a particular state could destroy homeschooling in that state. But the public school establishment... can afford to lose contest after contest...' (J.M. Smith 2006). Despite this awareness, the ability of the HSLDA/NCHE to channel its network's energies into individual state-level struggles has not yet been fully tested.

The importance of the National Home Education Research Institute, by contrast, derives from its ability to influence political discourse rather than its resource base. Largely the project of one individual researcher (Dr. Brian D. Ray), NHERI conducts and publishes largely favourable studies of homeschoolers' achievements. Despite its clear biases, it is one of the only sources for large-scale and longitudinal data on homeschooling activities. Part of NHERI's information monopoly derives from the air of privacy surrounding homeschooling, a factor which has always hindered research by 'outsiders'. But the dearth of official information is also a product of minimalist and fragmented regulation; NHERI's niche represents a fortuitous (or, perhaps, astute) articulation of its activities with the geographies of educational governance. State education authorities, even where homeschools are monitored, produce little in the way of systemic analysis. At the federal level, the lack of data may reflect a lack of regulatory authority. While the US Department of Education had a homeschooling specialist on staff until the mid-1990s, it now refers web visitors to NHERI for information. Currently, the agency tracks homeschoolers only as part of a larger, general-population survey of parents. NHERI, therefore, is a loud voice in homeschooling politics. Indeed, its studies are quoted frequently in media and policy reports.

Beyond the (conservative) political infrastructure represented by the HSLDA and NHERI, other spatial networks strengthen and deepen the homeschooling movement. Cultural and economic institutions, in particular, create bonds between dispersed practitioners. Specialized media, vendors, and movement celebrities are at the centre of this cultural-economic complex. For example, three bi-monthly publications, e.g. *Home Education Magazine*, *Practical Homeschooling*, and *Homeschooling Today*, reach tens of thousands of readers nationally. Their affiliate websites, meanwhile, log hundreds of thousands of hits each month. These media blend practical, inspirational, and political advice, and provide lively forums for discussion among practitioners. Elsewhere, a cadre of pedagogical experts, political commentators, and motivational speakers travel a circuit of well-attended homeschooling conventions and curriculum fairs. Even without their featured celebrities, these events provide a space for homeschoolers to construct and renew community, thus reproducing the basis for state- and national-level organizing. Finally, homeschoolers, with their US$700 million in annual education-related spending, are serviced by a growing number of specialist vendors (e.g. Walsh 2002). Established correspondence schools, Christian textbook publishers, and Internet distance education start-ups are among the many businesses catering to home educators. By providing coordinated, professionally-designed teaching materials, these firms have progressively eased the homeschooling experience for interested parents. In addition, they create shared experiences and vocabularies among practitioners. Homeschoolers, then, network beyond the home not just as political combatants, but as consumers and would-be teaching 'professionals'. This cultural-economic complex constitutes an enduring movement infrastructure, one that continues to involve homeschoolers between moments of political crisis (see Diamond 1997). As a particularly *spatial* networking strategy or moment of 'scale-jumping', it unites homeschoolers who might otherwise be divided by the specificities of localized regulation.

CONCLUSIONS: LOOKING FOR SPACE IN POLITICS AND POLICY

Because of its brevity, this mapping of homeschooling's institutionalization is necessarily suggestive. But the strategies outlined above indicate that an active manipulation of place and scale have contributed to the movement's development. To review, homeschoolers have justified relocating education by constructing the homeschool as a legitimate space for learning. Their work has further involved defining and defending the boundaries of this hybrid place against the state—a process that has selectively invoked a rich normative discourse linking home, family, and privacy in American political culture. Homeschoolers have further practised a politics of scale. By targeting the state-level in early struggles, and later developing institutions that can lithely traverse this still-fragmented regulatory terrain, homeschoolers have demonstrated a spatial flexibility. They have strategically worked within received geographies, transformed them, and built infrastructures that exceed or out-flank them.

On an empirical level, then, this review points to dimensions of the homeschooling story not yet addressed in the literature. Given the contested advance of educational privatization in the United States, this may constitute an important project in its own right. But as a theoretical contribution, this chapter argues for developing a constructivist spatial imagination in education policy scholarship more broadly. It is, at the core, a methodological intervention: if place and scale are part of political practice, they should become part of the conceptual repertoire used to analyze educational governance and policy development. Precisely *how* these might be incorporated will depend on the case in question. Indeed, after Miller (2000: 43), a geographical model of political mobilization 'has to be more of a sociospatial ontology—guiding the researcher toward crucial processes on which to focus—than a formal theory'. *Looking* for, or being open to strategic deployments and constructions of place and scale will contribute to fuller and more grounded accounts of educational restructuring.

The homeschooling case, of course, provides only limited examples of a spatial politics in action. In particular, its focus on a largely *deregulatory* moment of restructuring reduces the state and policy to secondary themes. In other arenas of educational politics, then, the geographies of state strategies—how the state organizes its capacities, deploys authority, and defines the objects of regulation—merit additional attention. Thus, recent political geographies of neoliberalism might be engaged to extend similar analyses of other privatization movements (e.g. Brenner 2004; Larner 2000; Peck 2003, 2004). More generally, a fully-realized spatial political analysis would engage more deeply with place and scale as structuring moments. While constructed, space is not infinitely malleable. A spatial politics is thus a complex process of articulation and iteration, involving the working out of competing spatial constructions and the layering of 'new' formations onto what came before.

In relation to this a spatial perspective on educational politics might be turned 'outwards' to considerable effect. In addition to explicating particular policy developments, it could be used to explore the spatial foundations for subsequent rounds of change. Homeschoolers' spatial constructions, for example, 'do work' for a variety of political projects. Thus, according to several authors, the homeschool is a place that supports a broader, conservative Protestant politics. It allows religious conservatives to 'demarcate and protect an alternate public space' (Bivins 2003: 106) and shelter children while preparing them for a life of spiritual warfare in the secular world (McDannell 1995). Viewed another way, the proliferation of homeschools may also affect the spatial constitution of institutional education. By moving education into the private home and participating in novel forms of educational delivery (e.g. online schools), homeschoolers facilitate the decoupling, both material and discursive, of state-sanctioned education from the public spaces of common schools.

Employed as a methodological approach, a constructivist spatial imaginary may thus enhance emerging narratives of educational restructuring. Rather than context, residual, or contingency, geography becomes both a stake and strategy in political mobilizations—a tool to initiate, stabilize, and contest policy change. While the case of US homeschooling reveals these moments in the context of a social movement, they may be found throughout the landscape of educational politics.

NOTES

1. In neither case do I argue that homeschooling advocates conceived of these strategies in explicitly spatial terms.
2. For additional applications, see collections and reviews in Herod and Wright (2002), Howitt (2003), and *Political Geography*, 16. 2.
3. Homeschooling advocates often invoke a lengthier historical lineage—from which *state* education represents a recent departure (e.g. Kirschner 1991). I find this claim problematic, less for its political overtones (which are clear enough) than for its silence around the changing *content* of education over time. Most of what constituted 'home education' in previous centuries bears little resemblance to even the most conservative forms of academic instruction offered in contemporary homeschools.
4. Many independent Christian schools (particularly in the US South) closed in the 1970s after losing their tax-exempt status. Authorities had charged the schools with practising racial discrimination.
5. For a more thorough analysis of the resonances between homeschooling and American political culture, see Stevens (2001). For a review of the popular equation of home and family, see Mallett (2004).
6. I use Cox's construct somewhat liberally here. In his original statement, a 'space of dependence' is defined in political-economic terms—material well-being and flows of value (e.g. jobs and investment) are prioritized (Cox 1998). The looser definition employed here positions the home as a space of dependence based on its centrality to the practice of homeschooling.
7. National groups formed by 'inclusive' homeschoolers have been hesitant to pursue explicitly political programs. For example, the National Home Education Network maintains a series of online forums for practitioners, but refrains from concerted organizing efforts. The loosely-organized National Homeschool Association was mired in procedural debates for much of its history, and dissolved in 2000. On divergent organizational histories of Christian and inclusive homeschoolers, see Stevens (2001).

REFERENCES

Allan, G. and Crow, G, (eds.) (1989) *Home and family: creating the domestic sphere*, London: Macmillan.

Apple, M. W. (2001) *Educating the 'right' way: markets, standards, god, and inequality*, New York: RoutledgeFalmer.

Bakker, I. (2003) 'Neo-liberal governance and the reprivatization of social reproduction: social provisioning and shifting gender orders', in I. Bakker and S. Gill (eds.), *Power, production and social reproduction: human in/security in the global political economy*, Basingstoke: Palgrave-Macmillan.

Bates, V. L. (1991) 'Lobbying for the lord: the new Christian right home-schooling movement and grassroots lobbying', *Review of Religious Research*, 33.1: 3–17.

Bielick, S., Chandler, K. and Broughman, S. P. (2001) *Homeschooling in the United States: 1999*, Washington, D.C.: National Center for Education Statistics, US Department of Education.

Bivins, J. C. (2003) *The fracture of good order: Christian antiliberalism and the challenge to American politics*, Chapel Hill: University of North Carolina Press.

Bosco, F. J. (2001) 'Place, space, networks, and the sustainability of collective action: the *Madres de Plaza de Mayo*', *Global Networks*, 1.4: 307–329.

Brenner, N. (2004) *New state spaces: urban governance and the rescaling of statehood*, Oxford: Oxford University Press.

Brown, M. P. (1997) *Replacing citizenship: AIDS activism and radical democracy*, New York: Guilford.

—— (2004) 'Between neoliberalism and cultural conservatism: spatial divisions and multiplications of hospice labor in the United States', *Gender, Place and Culture*, 11.1: 67–82.

Carper, J. C. (2000) 'Pluralism to establishment to dissent: the religious and educational context of home schooling', *Peabody Journal of Education*, 75.1–2: 8–19.

Carper, J. C. and Ray, B. D. (2002) 'Religion, schooling, and home education: past and present', in E.L. Blumhofer (ed.), *Religion, education and the American experience: reflections of religion and American public life*, Tuscaloosa, AL: University of Alabama Press.

Cibulka, J. G. (1991) 'State regulation of home schooling: a policy analysis', in J. Van Galen and M. A. Pitman (eds.), *Home schooling: political, historical, and pedagogical perspectives*, Norwood, NJ: Ablex.

Cox, K. R. (1998) 'Spaces of dependence, spaces of engagement and the politics of scale, or: looking for local politics', *Political Geography*, 17.1: 1–23.

Delaney, D. and Leitner, H. (1997) 'The political construction of scale', *Political geography*, 16.2: 93–97.

Diamond, S. (1997) *Not by politics alone: the enduring influence of the Christian right*, New York: Guilford.

Gallup, A. M. (1985) 'The 17th annual Gallup poll of the public's attitudes toward the public schools', *Phi Delta Kappan*, 67.1: 35–47.

Harvey, D. (1982) *The limits to capital*, Oxford: Blackwell.

—— (1985) 'The geopolitics of capitalism', in D. Gregory and J. Urry (eds.), *Social relations and spatial structures*, London: Macmillan.

Herod, A. (1991) 'The production of scale in United States labor relations', *Area*, 23.1: 82–88.

Herod, A. and Wright, M. (eds.) (2002) *Geographies of power: placing scale*, Malden, MA: Blackwell.

Home School Legal Defense Association (1998) *Marking the milestones*, accessed 27 February, 2006, http://www.hslda.org.about/history/default.asp

hooks, b. (1990) *Yearning: race, gender, and cultural politics*, Boston: South End Press.

Howitt, R. (2003) 'Scale', in J. Agnew, K. Mitchell and G. Toal (eds.), *A companion to political geography*, Malden: Blackwell.

Huerta, L.A. and Gonzalez, M. (2004) *Cyber and home school charter schools: how states are defining new forms of public schools*, National Center for the Study of Privatization in Education, Occasional Paper no. 61.

Jones, K. (1998) 'Scale as epistemology', *Political Geography*, 17.1: 25–28.

Keith, M. and Pile, S. (eds.) (1993) *Place and the politics of identity*, London: Routledge.

Kirschner, J. (1991) 'The shifting roles of family and school as educator: a historical perspective', in J. Van Galen and M. A. Pitman (eds.), *Home schooling: political, historical, and pedagogical perspectives*, Norwood, NJ: Ablex.

Klicka, C. J. (2002) 'Charter schools: the price is too high', *Home school court report,* accessed 27 February, 2006, 18.1. http://www.hslda.org/courtreport//V18N1/V18N101.asp.

Larner, W. (2000) 'Neo-liberalism: policy, ideology, governmentality', *Studies in Political Economy*, 63: 5–25.

Lefebvre, H. (1991) *The production of space*, London: Blackwell.

Leitner, H., Pavlik, C. and Sheppard, E. (2002) 'Networks, governance, and the politics of scale: inter-urban networks and the European Union', in A. Herod and M. Wright (eds.), *Geographies of power: placing scale*, Oxford: Blackwell.

Lines, P. M. (1998) *Homeschoolers: estimating numbers and growth*, Washington, D.C.: National Institute on Student Achievement, Curriculum, and Assessment, Office of Education Research and Improvement, US Department of Education.

Lubienski, C. (2000) 'Whither the common good? A critique of home schooling', *Peabody Journal of Education,* 75.1–2: 207–232.

McDannell, C. (1995) 'Creating the Christian home: home schooling in contemporary America', in D. Chidester and E. T. Linenthal (eds.), *American sacred space*, Bloomington, IN: Indiana University Press.

Mallett, S. (2004) 'Understanding home: a critical review of the literature', *Sociological Review,* 52.1: 62–89.

Marston, S.A. (2000) 'The social construction of scale', *Progress in Human Geography,* 24.2: 219–42.

Massey, D. (1994) *Space, place and gender*, Minneapolis: University of Minnesota Press.

—— (2005) *For space,* London: Sage.

Mayberry, M. (1988) 'Characteristics and attitudes of families who home school', *Education and Urban Society,* 21.1: 32–41.

Miller, B. (1994) 'Political empowerment, local-central state relations and geographically-shifting political opportunity structures', *Political Geography*, 13.5: 393–406.

—— (1997) 'Political action and the geography of defense investment: geographical scale and the representation of the Massachusetts Miracle', *Political Geography,* 16.2: 171–185.

—— (2000) *Geography and social movements: comparing anti-nuclear activism in the Boston area*, Minneapolis: University of Minnesota Press.

Munro, M. and Madigan, R. (1999) 'Negotiating space in the family home', in I. Cieraad (ed.), *At home: an anthropology of domestic space*, Syracuse, NY: Syracuse University Press.

National Center for Education Statistics (2004) *1.1 million homeschooled students in the United States in 2003*, Washington, D.C.: US Department of Education.

Peck, J. (2002) 'Political economies of scale: fast policy, interscalar relations, and neoliberal workfare', *Economic Geography*, 78.3: 331–360.

—— (2003) 'Geography and public policy: mapping the penal state', *Progress in Human Geography*, 27.2: 222–232.

—— (2004) 'Geography and public policy: constructions of neoliberalism', *Progress in Human Geography*, 28.3: 392–405.

Reich, R. (2002) 'The civic perils of homeschooling', *Educational Leadership,* 59.7: 56–59.

Riegel, S. (2001) 'The homeschooling movement and the struggle for democratic education', *Studies in Political Economy*, 65: 91–116.

Robertson, S. and Dale, R. (2002) 'Local states of emergency: the contradictions of neo-liberal governance in education in New Zealand', *British Journal of Sociology of Education*, 23.3: 463–482.

Rose, L. C. and A. M. Gallup (2001) 'The 33rd annual Phi Delta Kappa/Gallup poll of the public's attitudes toward the public schools', *Phi Delta Kappan*, 83.1: 41–58.

Routledge, P. (1997) 'A spatiality of resistances: theory and practice in Nepal's Revolution of 1990', in S. Pile and M. Keith (eds.), *Geographies of resistance*, London: Routledge.

Rykwert, J. (1991) 'The idea of a home: a kind of space', *Social Research*, 58.1: 51–62.

Saunders, P. and Williams, P. (1988) 'The constitution of the home: towards a research agenda', *Housing Studies*, 3.2: 81–93.

Smith, C. and Sikkink, D. (1999) 'Is private schooling privatizing?' *First Things*, 92 (April): 16–20.

Smith, J. M. (2006) 'Remaining vigilant protects gains', *Washington Times*, 30 January.

Smith, N. (1984) *Uneven development: nature, capital and the production of space*, Oxford: Blackwell.

——— (1992) 'Contours of a spatialized politics: homeless vehicles and the production of geographical space', *Social Text*, 33: 54–81.

——— (1993) 'Homeless/global: scaling places', in J. Bird, B. Curtis, T. Putnam, G. Robertson and L. Tickner (eds.), *Mapping the futures: local cultures, global change*, New York: Routledge.

Soja, E. (1980) 'The sociospatial dialectic', *Annals of the Association of American Geographers*, 70: 207–225.

——— (1989) *Postmodern geographies*, Oxford: Blackwell.

Staeheli, L.A. (1994) 'Empowering political struggle: spaces and scales of resistance', *Political Geography*, 13.5: 387–391.

——— (1996) 'Publicity, privacy, and women's political action', *Environment and Planning D: Society and Space*, 14(5): 601–619.

——— (2003) 'Place', in J. Agnew, K. Mitchell and G. Toal (eds.), *A companion to political geography*, Malden, MA: Blackwell.

Stevens, M.L. (2001) *Kingdom of children: culture and controversy in the homeschooling movement*, Princeton, NJ: Princeton University Press.

Swyngedouw, E. (1997) 'Neither global nor local: "glocalization" and the politics of scale', in K. Cox (ed.), *Spaces of globalization: reasserting the power of the local*, New York: Guilford.

Taylor, P. (1982) 'A materialist framework for political geography', *Transactions of the Institute of British Geographers*, 7.1: 15–34.

Thrift, N. J. (1996) *Spatial formations*, London: Sage.

——— (1999) 'Steps to an ecology of place', in D. Massey, J. Allen and P. Sarre (eds.), *Human geography today*, London: Polity.

Tyler, Z. (2003) 'State organizations: making our voices heard', *Home School Court Report*, 19.1, accessed 27 February 2006, http://www.hslda.org/courtreport/V19N1/V19N101.asp.

Van Galen, J. (1991) 'Ideologues and pedagogues: parents who teach their children at home', in J. Van Galen and M.A. Pitman (eds.), *Home schooling: political, historical, and pedagogical perspectives*, Norwood, NJ: Ablex.

Walsh, M. (2002) 'Home school enrollment surge fuels 'cottage' industry', *Education Week*, 5 June.

'We Stand for Homeschooling', accessed 27 February 2006, http://westandforhomeschooling.org.

3 Mobilizing space discourses

Politics and educational policy change

Kalervo N. Gulson

> ...thinking the spatial in a particular way can shake up the manner in which certain political questions are formulated, can contribute to political arguments already underway, and—most deeply—can be an essential element in the imaginative structure which enable[s] in the first place an opening up to the very sphere of the political.
>
> Massey (2005: 9)

How might a poststructural notion of space add to the opening up of the 'sphere of the political' within educational policy research? Henri Lefebvre (1991) posited a 'politics of space' around capital's implication in the production of space; and the spatial aspects of politics have been raised in such areas as spatial resistance and political struggle (see Keith 1997; Massey 1994); while spatial aspects of policy have been identified in relation to a variety of issues such as mobility (see Jensen and Richardson 2004; Strüver 2004). In relation to a European Union 'spatial policy,' Tim Richardson and Ole Jensen's (2003) work 'focuses on how words and actions frame and represent spaces on the basis of certain relations between power and rationality' (p. 8). In a similar vein, the focus of this chapter is using a poststructural spatial analytic to explore the political mobilization of space discourses during educational policy change. This is illustrated by examining aspects of the *Building the Future* policy initiative in inner Sydney, Australia, and the subsequent New South Wales[1] (NSW) Parliamentary Inquiry into this policy initiative.

In 2001 the NSW Department of Education and Training (DET) released the *Building the Future: an education plan for inner Sydney* (DET 2001a) policy that focused on restructuring state schooling in inner Sydney. After a period of consultation a second policy statement was released, the *Building the Future: consultation report* (DET 2001b). These policy documents outlined how the DET aimed to 'revitalize [inner] city schools so that they offer better educational opportunities and are more attractive in the future' (DET 2001b: 1). This 'revitalization' was to be underpinned by structural change through the closure and amalgamation of a number of primary

(Years K–6) and secondary schools (Years 7–12) (see Blackmore 2002; Gulson 2004). Schools that were proposed to close and amalgamate in the inner city area of South Sydney included Erskineville, Redfern, Waterloo, and Alexandria primary schools and Cleveland Street High School.

The DET justified closing smaller schools on the basis that steadily falling enrollments over a period of 20 years had resulted in a lack of both curriculum choice and extra-curricular activities. However, during 2001 and 2002 the activities of lobby groups that were school- and community-based, such as the utilization of public meetings and media coverage including letters to the Sydney daily papers, brought political pressure to bear on the NSW Labor Government that faced an upcoming election in 2003. In early 2002, within this highly charged political environment, the opposition parties in the NSW Parliament's Upper House under the aegis of the New South Wales Parliament Legislative Council: General Purpose Standing Committee No 1 (Legislative Council) held the *Inquiry into proposed closure and restructuring of inner Sydney schools*, with a series of public proceedings held during May and June 2002. The transcripts of these proceedings were made available to the public on the NSW Parliament website (Legislative Council 2002b, 2002c, 2002d), and through the Inquiry's final report (Legislative Council 2002a). These transcripts are used as illustrative data in this chapter.

In this chapter I first locate the uptake of poststructuralism within education policy studies, specifically in terms of policy sociology; second, I link poststructuralism to 'relational space' in education policy sociology; and, in the last section of the chapter, using the Inquiry transcripts, I apply a poststructural spatial analytic to aspects of the Legislative Council Inquiry. I identify how spatial knowledges were politically mobilized as legitimate and/or illegitimate, particularly around notions of 'scientificity', during the processes and practices of inner city educational policy change.

THE ROLE OF POSTSTRUCTURALISM IN EDUCATION POLICY SOCIOLOGY

Within education policy sociology (see Ball 1990, 1994b; Ozga 1987, 1990), policy analyses using poststructuralist analytics have identified the often simultaneous practices of consensus and contestation (see, for example, Ball 1994a, 2000; Kenway 1990; Olssen 2004; Youdell 2004). However, it is important to acknowledge that poststructuralism may not be 'method' as such, but rather there may be poststructural forms of 'criticism', such as deconstruction (Cherryholmes 1988 cited in Peters and Burbules 2004).[2] In this chapter, when I refer to the 'poststructural' or 'poststructuralism' I am using it as shorthand for an approach that draws on theorists such as Michel Foucault (despite his refusal to be labelled as such), and concepts such as discourse, broadly speaking. I recognize that this may inadvertently do the type of totalising work that is challenged by poststructural approaches. Nonetheless, I accept that poststructural forms of criticism may 'produce an "incredulity towards metanarratives," to dissemble the structures, the "moves" and strategies of official discourse' (Peters and Burbules 2004: 5). This dissembling is what Bronwyn Davies (2004) sees as disrupting 'truths'. Thus, a possible aim of a poststructuralist analysis or criticism is 'not to expose the hidden truth in all its simplicity, but to *disrupt that which is taken as stable/ unquestionable truth*' (Davies 2004: 7, original emphasis). As well as disruptions, poststructuralist approaches such as deconstruction can also provide openings. 'The kind of opening which such work attempts is that of rupture—of interruption and disruption—in the (uncertain) hope that this will generate possibilities for things to happen that are closed off by the epistemologies of certainty' (Stronach and MacLure 1997: 5).

The possibilities, or openings, that this approach may create in educational policy sociology can be illustrated by examining aspects of the Legislative Council's Inquiry into school closures. In this illustration Stephen Ball's (1993) two distinct, but interrelated, notions of 'policy as text' and 'policy as discourse' are an impetus. Ball asserts that policy is not either 'policy as text' *or* 'policy as discourse', rather:

> they are 'implicit in each other'. Policy discourses...produce frameworks of sense and obviousness with which policy is thought, talked and written about. Policy texts are set within these frameworks, which constrain but never determine all of the possibilities for action. (Ball 2006: 44)

These discursive notions build on Richard Bowe, Stephen Ball, and Anne Gold's (1992) 'contexts of policy' to explicitly engage with policy as contestation. In relation to *policy as text*, Ball (1993) notes there is a sense of incompleteness, that '[a] policy is both contested and changing, always in a state of "becoming," of "was" and "never was" and "not quite"' (p. 11). This conceptualization speaks against the linearity of policy and has resonance with iterative models of the policy cycle (see, for example, Howlett and Ramesh 1995).

> There is ad hocery, negotiation and serendipity within the state, within the policy formulation process.... [However] only certain influences and agendas are recognized as legitimate, only certain voices are heard at any one point in time within the commonsense of policy. The point is that quibbling and dissensus still occur within the babble of 'legitimate' voices and sometimes the effects of quibbling and dissensus result in a blurring of meanings within texts and in public confusion and a dissemination of doubt. (Ball 2006: 45)

In the transcripts of the Legislative Council Inquiry it is apparent that there were significant disagreements between individuals and groups over the directions set out in the *Building the Future* policy documents. During the Inquiry many of the submissions alluded to the variety of inner city 'community' uses for the schools under the threat of closure. These submissions were made to four hearings held over the course of two months in mid-2002. The Inquiry's terms of reference appeared to be premised on the principle of equal access to all interested parties, but in practice the Inquiry appeared to instigate a principle of differentiated access. The DET's representatives fronted the Inquiry for two hearings in which they gave the majority of evidence, whereas other groups and individuals only gave evidence at one hearing each. Additionally, the DET representatives were the only ones who elected to present information 'in camera' to the Inquiry, thereby preventing certain information from being recorded for the public record. This practice resonates with a notion of power that, following Foucault

> seeks to show how relations of 'agency' and 'structure' have been constituted discursively, how agency is denied to some and given to others, how structures could be said to have determined some things and not others. The focus is on how certain forms of representation are constituted rather than upon the 'truth' or 'falsity' of the representations themselves. (Clegg 1989: 158)

Power is also a consideration in terms of *policy as discourse*, '[d]iscourses are about what can be said, and thought, but also about who can speak, when, where and with what authority' (Ball 1993: 14). What is central here is that 'power is very much implicated in the production of knowledge. What can be said? Who can say it?' (St Pierre 2000: 496). And where can they say it? As such, policy as discourse recognizes 'regimes of truth' in which:

> [t]here is a battle 'for truth,' or at least 'around truth'—it being understood once again that by truth I do not mean 'the ensemble of rules according to which the true and the false are separated and specific effects of power attached to the true', it being understood also that it's not a matter of a battle 'on behalf' of the truth, but of a battle about the status of truth and the economic and political role it plays. (Foucault 1980: 132)

During the Inquiry the 'battle about the status of truth' about inner city schools was fought over which 'representation' of schools had more 'claim' to the 'truth'. The DET's version highlighted lower enrolments and struggling schools, and mobilized an epistemological armoury to fight the power/knowledge battle, supported by demographers and enrolment data that pointed to less curricular and extra-curricular diversity. The school and community advocates' versions had recourse to the centrality of schools in inner city locations in relation to not only education, but also social interaction. I return to these spatial aspects in the last section of the chapter.

It is obvious, as has been argued previously (e.g. Ball 1994a), that contestation, and discourses of certainty, can be explored, and disrupted, by a poststructural analytic employed within educational policy sociology. However, I wish to also acknowledge Ball's (2006) concern that '[p]olicy research lacks a sense of "place"; either in not locating policies in any framework that extends beyond the national level, or in not accounting for or conveying a sense of the locality in the analyses of policy realisation' (p. 19). It would appear that the type of analysis called for by Ball requires additional 'tools' than those outlined in the poststructural policy analysis above. Here, Ball's (1993) concern is pertinent, where:

> [t]he complexity and scope of policy analysis...precludes the possibility of successful single theory explanations. What we need in policy analysis is a toolbox of diverse concepts and theories. (p. 10)

Therefore, what I would like to suggest is that identifying the mobilization of space discourses can contribute to the 'toolbox' of a poststructural policy sociology analysis. This is premised on what may be seen as the 'spatial' aspect lacking in much social science research; an aspect that may usefully be considered in addition to temporal concerns (see, for example, Soja 1996). However, when the spatial aspect is invoked, 'space' as an analytical tool in the social sciences is, in many instances, applied without adequate definitional rigour (as argued by, for example Crang and Thrift 2000; Massey 1993a). Following these concerns, it appears that the initial step may be to identify possible space discourses available to be mobilized.

A SPATIAL ANALYTIC FOR EDUCATION POLICY SOCIOLOGY

During the twentieth century space had been 'reasserted' in social theory (see Soja 1989), and one field within which this 'reassertion' manifested was education policy sociology (see Ball, Maguire, and Macrae 1998, 2000; Gulson 2005, 2006; Robertson 2002). This 'reassertion' in social theory prompted Neil Smith (2004) to suggest that:

> [t]here is a crucial question of the extent to which this 'spatial turn' has been more than skin deep....Put most crudely, perhaps, why space? Why *should* our analysis of social difference and political possibility be rewritten in the language of space? (Smith 2004: 13, original emphasis)

For education policy sociology the 'language of space' may allow for explorations of education policy changes that are increasingly concerned with both the interconnectedness of social policy trajectories (e.g. S. Taylor, Rizvi, Lingard, and Henry 1997) and with the enactment of 'localized' educational policy changes and constructs such as the education market (e.g. Ball, Bowe, and Gewirtz 1995; C. Taylor 2001). These explorations seem, therefore, to require viewing policy as both spatial and temporal, or as Doreen Massey (1993a) argues, 'spatiality and temporality are different from each other but neither can be conceptualized as the absence of the other' (p. 155). In a sense, while the historical aspects of policy have been well defined, the definition of the spatial in fields such as education has been less precise,[3] though other social science disciplines such as social and cultural geography and sociology have engaged in vigorous debate.

Since the 1960s, a key question for the social sciences has been whether there is a link between society and space, and if so, what are the characteristics of this link? In the 1960s spatial metaphors and language were used to identify social patterns, and in this approach:

> [t]he rhetoric of spatial fetishism...targeted a kind of 'identity thinking' in which people are assimilated to the spaces they occupy and are viewed as attributes of these spaces, and spaces are ascribed the qualities of their human residents and appear to take on a life of their own. (Collinge 2005: 191)

This position was challenged by the assertion of social determinism; a sociological rejection of 'spatial fetishism' evident in such work as Manuel Castells (1977 cited in Collinge 2005), in which the social characteristics of cities are determined by factors such as class, gender, 'race', and not the physical environment. This social determinism was in tension with the integration of society and space offered by Lefebvre's (1991) socio-spatial dialectic. For Lefebvre (1991) space as a producer of, and produced by social relations, requires that the interactions of people are shaped in some way by the spatial surroundings and that social relations may shape spatial surrounds. However, Chris Collinge (2005) contends that arguments over social determinism, spatial determinism, and dialectic propositions 'took place within a frame of reference that generally relied upon a dualistic distinction between society and space, now for the most part understood dialectically as involving the identity of opposites' (p. 192).

In a similar direction to the sociological arguments over the dialectic of structure and agency, some geographers moved towards a poststructuralist emphasis that focussed, not on oppositions, but rather on indeterminancy, and frames of reference in which 'the [real] neither yields meaning nor assumes it except in a movement of differential referring' (Derrida 1988: 148 cited in Collinge 2005: 193). Much of this work occurred through what might be termed 'third space' (e.g. Soja 1996). Paul Cloke and Ron

Johnston (2005), drawing on Steve Pile (1994: 273), outline how this works to permit:

> a positioning of socially constructed binaries within contexts of dissimilarity; an appreciation of how the values established by power-laden binaries can be decoded and challenged; a recognition of the potential which exists beyond the interconnecting power of binaries; and an understanding of the fluidity of space when thinking through 'the ways in which the fabric of this third space is continually fragmented, fractured, incomplete, uncertain and the site of struggles for meanings and representation'. (Cloke and Johnston 2005: 19)

I am hesitant to agree with the possibility of something 'beyond' binaries, though I am happy to ascribe to the project of unfixing and destabilising binaries pursued by Cloke and Johnston (2005). It is through this unfixing and destabilizing, in the realm of openings and possibilities,[4] that I see a confluence of a poststructural policy sociology and a poststructural geography.

A POSTSTRUCTURAL GEOGRAPHY IN EDUCATIONAL POLICY SOCIOLOGY?

There is resonance between poststructuralist criticism as applied in education policy sociology (see Peters and Humes 2003: 111–113) and in social and cultural geography (see Doel 1999, 2000, 2004a, 2004b), particularly around practice and contestation. As Derek Gregory (1994) argues in reference to taking up approaches from feminism, poststructuralism, and postcolonialism 'any critical human geography must attend to the ways in which meanings are spun around the *topoi* of different lifeworlds, threaded into social practices and woven into relations of power' (p. 76) and that human geography is 'an irredeemably situated, positioned system of knowledge' (p. 76).

Thus, for example, in reference to geography and poststructuralism, Marcus Doel asserts that:

> poststructuralism is always already spatial: that it attends from the off to the 'difference that space makes'. It is the event of space, of spacing, that deconstructs. Speaking always as geographers, poststructuralists strain to become sensitive to what it is to space—to what happens when space takes space. (Doel 1999: 10)

Deconstruction in geography challenges the possibility of coherent, complete, and stable representation/s (see Doel 2005). It shifts the terrain to what Doel posits as the possibilities of 'becoming-otherwise' underpinning a poststructuralist geography(ies), where '[g]eography is an act, an event, a happening' (Doel 1999: 11). Whilst perhaps agreeing with the sentiment, Massey rejoins that:

> rather...poststructuralism *could* very easily be spatial.... But, as Derrida himself points out, for deconstruction to live, and particularly when it is being transported into new areas, it will need to be transformed.... [T]he sympathetic trick is to work within but to emerge, maybe, with something appropriately different. (Massey 2005: 50, original emphasis)

What I think Massey is getting at here, perhaps obviously, is that Doel, in claiming that 'poststructuralism is always already spatial', is closing down what he claims to be opening up; it may be that poststructuralism is not *a priori* spatial, but rather that a poststructural analytic results in possibilities of thinking through, and of, different types of space—spaces of relations, multiplicity and politics (Massey 2005). A poststructuralist position recognizes the mobilization of different types of space discourses, but has an emphasis on 'relational space' that is, following particularly Massey (1993b), concerned with power relations as both constraining and enabling (Murdock 2006). Jonathan Murdock (2006 following Thrift 2004) suggests that poststructural ideas are productive in terms of possibility or 'potentials', where there is an intersection and combination of social and spatial relations such that 'space is practised and performed in the same way that social identity and belonging are practised and performed' (Murdock 2006: 17). The notion of space as performative is raised by Gregory (2004), who initially draws on a theatrical motif of performance, and then extends this to encompass notions of performativity, notably in relation to the work of Judith Butler. Gregory follows Foucault's (1972) notion of discourse 'as practices that systematically form the objects of which they speak' (p. 49), to posit imaginative geographies as 'not only accumulations of time, sedimentations of successive histories; they are also *performances of space*' (Gregory 2004: 18–19, original emphasis), and suggests that 'performance creates a space in which it is possible for "newness" to enter the world' (p. 19). Implicit in the notions of performance and performativity, again in reference to Butler in particular (e.g. 1997) are power and interrelations; and as Nicky Gregson and Gillian Rose (2000) explicate:

> Our argument is that performance—what individual subjects do, say, 'act-out'—and performativity—the citational practices which reproduce and/or subvert discourse and which enable and discipline subjects and their performances—are intrinsically connected, through saturation of performers with power.... Space too needs to be thought of as brought into being through performances and as a performative articulation of power. (Gregson and Rose 2000: 434)

However, a focus on 'relational space' could be posited as an attempt to supplant the 'relationality' of 'time', and hence manifest Foucault's (1986) observation, talking in the mid-twentieth century, that whereas the nineteenth century had seen a focus on 'history', or time, 'the present epoch will perhaps above all be the epoch of space' (Foucault 1986: 22). Despite this possibility, there is again an alternative in destabilizing boundaries—inherent in Massey's (1993a) conception of 'space/time' in which both time and space are relational and which 'insist[s] on the inseparability of time and space, on their joint constitution through the interrelations between phenomena; on the necessity of thinking in terms of space-time' (p. 159). And in more recent work Massey (2005) suggests that:

> [c]onceiving of space as a static slice through time, as representation, as a closed system and so forth are all ways of taming it....If it is to be open to a future of the new then space cannot be equated with the closures and horizontalities of representation. More generally, if time is to be open then space must be open too. Conceptualising space as open, multiple and relational, unfinished and always becoming, is a prerequisite for history to be open and thus a prerequisite, too, for the possibility of politics. (Massey 2005: 59)

It is not only the binary of space and time that can be opened up by an emphasis on multiplicity and relationality. Educational policy change similarly may be considered to be practised and performed in particular places; as Ball (2006) notes a sense of 'place' is an important aspect of understanding policy change. However, when constructing a spatial analytic 'place' can be positioned in a binary with space; where it is place *or* space; place *over* space; space *over* place (for an elegantly simple discussion of the place/space binary, see Agnew 2005). Taking seriously the unfixing of binaries means multiplicity and dislocation are recognized as significant aspects of both place and space where:

> the emphasis is on places as sites in the flow of social relations. Place is seen as constituted out of space-spanning relationships, place specific social forms and a sense of place associated with the relative well-being, disruption and experience of living somewhere. (Agnew 2005: 91)

So, for example, Massey's (2005) notion of place as an 'event' positions place and space as interrelated rather than oppositional. And, as Doel (1999) asserts, 'place is nothing if it is not *in process*' (p. 7). Furthermore, place as a part of social relations can be 'imagined as articulated moments in networks of social relations and understandings' (Massey 1993b: 66). As I have noted elsewhere, place as an articulated moment of social space has resonance with educational policy change as involving conflict, discursive manoeuvring, and compromise (Gulson 2006). Therefore, building on the resonance between place, policy, and space, what might a poststructural spatial analytic look like in a policy sociology analysis?

MOBILIZING SPACE DISCOURSES: A POSTSTRUCTURAL SPATIAL ANALYTIC AND POLICY CHANGE IN THE INNER CITY

As I have noted above (and following, for example, Ball 1993), 'regimes of truth' can operate to open up and close down, enable and constrain within policy as contestation. Here I posit that the link between poststructuralism and relational space may usefully add to understanding politics and policy contestation, where the mobilization of space discourses may be identified in the practices surrounding educational policy change. Specifically, using data as illustrative and exploratory, I will examine the mobilization of 'absolute space' and 'relational space' during the Legislative Council Inquiry into school closures in inner Sydney. 'Relational space' in this section posits interrelations as a way of making sense of policy change around particular constitutions of spaces and places in the inner city. This is a mobilization of relational space as content, or the relations within space. This is to distinguish it from the notion of relational space as a floating signifier, a concept with a focus on multiplicity and becoming. The way that relational space is mobilized within the Inquiry is closely related to Murdock's (2006) conception of:

> [t]he relational making of space [as] both a consensual and contested process. 'Consensual' because relations are usually made out of agreements or alignments between two or more entities; 'contested' because the construction of one set of relations may involve both the exclusion of some entities (and their relations) as well as the forcible enrolment of others. In short, relational space is a 'power-filled' space in which some alignments come to dominate, at least for a period of time, while others come to be dominat*ed*. So while multiple sets of relations may well co-exist, there is likely to be some competition between these relations over the composition of particular spaces and places. (Murdock 2006: 20, original emphasis)

In a sense, the converse of relational space is 'absolute space' or 'abstract space' that had dominated social theory.[5] This is space as understood through Euclidean geometry, as measurable and able to be mapped, with a focus on space as empty, 'a conception of space as a field, container, a co-ordinate system of discrete and mutually exclusive locations' (Smith and Katz 1993: 75). Space is something to be filled, with numbers, people, objects, and once filled it is static and measurable. This is a particular type of spatial 'science' in which social processes are represented by, and in, mathematical terms, which is not to say that social processes are unable to be represented in such a way, but rather there are limits to the utility of this representation (Crang and Thrift 2000: 2; Gregory 1994: 90). Additionally, while 'abstract space' is closely associated with geography's 'quantitative revolution' during the 1950s and 1960s, Smith (2004) suggests that, in the late twentieth century, there was a reassertion of the abstraction of space that 'has led to a certain desocialization of geography, most evident in the equations and algorithms of positivist geography and the abstractions of [Geographic Information Systems] technologies that blur the always-contested translations linking representation and reality' (p. 26).

What is striking here is that absolute space, it seems, is also discursively produced; as such it may be mobilized in policy and in terms of politics. There is a lineage here with Lefebvre's (1991) notion of 'abstract space' as the space of capitalism implicated in the production of space. For example, Lefebvre (1991) outlined how space as produced by social relations is seen in the realm of *representations of space (conceived space)* as 'conceptualized space[s]' in which people such as 'scientists, planners, urbanists, technocratic subdividers...all...identify what is lived and what is perceived with what is conceived' (Soja 1996: 66–67). Edward Soja suggests that this is about order, and '[s]uch order is constituted via control over knowledge, signs, and codes: over the means of deciphering spatial practice and hence over the production of spatial knowledge' (Soja 1996: 67).

It is the invocation of these space discourses that I am interested in, particularly how, in policy discourses circulating around inner city educational policy change, these notions of 'relational' and 'absolute' space are mobilized for political purposes.

THE MOBILIZATION OF ABSOLUTE AND RELATIONAL SPACE(S) IN INNER SYDNEY

The *Building the Future* policy change had a central focus on demographics as a justification for structural change. For example, in the *Building the Future* documents, the DET argued that there was a significant movement out of the inner city area claiming that:

- there are fewer children in modern families
- families are moving from inner city to outer suburban areas for affordable housing (DET 2001a: 4).

During the Legislative Council Inquiry, the DET responded to challenges to closure through recourse to, and mobilization of, particular spatial discourses. These claims employed demographics and enrolment data that were used to construct a sense of inevitability about the decisions to close inner city schools. These claims also constituted the inner city in a particular way, as a city irreversibly emptying of children. There was a specific type of space that was given primacy; the primacy of 'absolute space', which was legitimated through recourse to demographic representations. The inner city, and inner city schools, was, and were, prior to measurement, 'empty spaces', which then became the spaces of demographics. While the DET mobilized demographics as a way of justifying closing schools, those opposed to the closures mobilized similar, but oppositional, discourses around 'absolute space'. For example, one position taken by the Erskineville Parents and Citizens Association (P & C) was to challenge the accuracy of the DET's demographic conclusions. Ms Mulvey, the president of the Erskineville Primary School Parents and Citizens Association, asserted that:

> The [DET's] case for closure of Erskineville Public School rests on decline of enrolments and outdated ABS data. They focus on the past. The Erskineville Parents and Citizens Association case that we have documented many times to them rests on the most current demographics. It includes (a) a survey of child numbers at local child care centres; (b) figures from local government, and (c) the most recent ABS data. (Legislative Council 2002d: 2)

Here there is an assumption that measurement can be 'up-to-date,' despite the idea that demographics, even over time, may be a representation that 'em-places' and thus is always 'in the past'. Thus, Ms Mulvey's position appears to represent the preciseness of demographics, while being employed to cast doubt on the DET's projections. However, Ms Mulvey's case was undermined by another submission that was, ostensibly, to support the position of the Erskineville P & C. This submission by Dr Phibbs, a demographer from the University of Sydney, conceptually confronted the notion that demographics, in any form, were precise.

> I do population forecasts for local government areas. I had previously completed a population forecast of South Sydney, which I think a colleague at work knew, and one of her children went to Erskineville [Primary School], so she asked me to look at the demographic report that was part of the [DET's] work....
>
> When I read it I was a little alarmed because I think it gave the impression that forecasting is a more precise science than it actually is. I cannot remember the last time I saw a forecast that was correct. There are so many risks involved in forecasting that basically a better way to think of them is as estimates. (Legislative Council 2002d: 6)

There are further examples found throughout the Inquiry transcripts that mobilize a particular construction of demographic knowledge. This testimony by demographers includes:

> Therefore...the second [demographic] report...I have to say I think that it is slighted a bit favourably towards Hunters Hill High School, *but it is a scientific piece of work*.... The other [demographic] reports by Yusuf et al, and also by Phibbs, are *scientific pieces of work*. (Legislative Council 2002c: 41, my emphasis)

> I think, given a number of the assumptions, that both sets of [demographic] reports are fair. *They are both scientific inquiries*. (Legislative Council, 2002c: 42, my emphasis)

Therefore, it seems, the reports can be challenged in terms of the claims made, and even in terms of the 'predictive' function of the demographic forecasting but the methodologies, the 'scientificity', and subsequently the legitimacy of these claims in the policy debate is unquestionable. This reflects the mobilization of particular spatial discourses associated with 'absolute space', where the political ramifications of spatialized 'regimes of truth' are evident around the recourse to 'scientificity'. As Petersen (2004), following Foucault, suggests, 'scientificity' may be conceptualized:

> *as something that can be established*. Even as something we can strain to establish. By investing itself in recognized 'scientific' discourses and practices, an act, articulation, or body, might therefore then come to be recognized as 'scientific'. In other words this formulation 'straining to establish the scientificity,' theorizes that something that is recognized as not, as *outside* the category, might in various ways be worked at, re-inscribed, to become recognized as *inside* the category. (Petersen 2004: 183, original emphasis)

The use of this type of demographic data reinforces an aspiration within policy making, and analysis, to the 'predictive'. It is the language of 'absolute space' that presents a level of preciseness. Thus, the DET supports the policy directions outlined in *Building the Future* through the political mobilization of demographics, including falling enrolment numbers, whilst recourse to 'abstract space' simultaneously, and paradoxically, denies the politics of a policy process.

Alternatively, discourses mobilized around relational space were evident in the transcripts from the Legislative Council Inquiry. For example, a number of submissions raised the issue of young children travelling to a new P-12 school, to be created from the closed inner city schools. Those who opposed the closure of schools emphasized the danger of children crossing busy city roads and travelling across suburbs with inadequate public transport. The DET's submissions first dismissed the concerns around travelling across roads, and second posited solutions based on providing transport to the P-12 school. Dr Laughlin, Deputy Director of DET, outlined that:

> [i]n response to the claim that students would be at risk or that transport would be problematic, we would like to bring these following matters to...attention. In this part of the city the primary schools are in fact only about 600 to 700 metres apart. They are very close. Many students are already travelling across busy roads to Erskineville and further than they would if they attended the new [P-12 school].... We have also had five meetings so far with Sydney buses about transport to [P-12 school], and we believe we can have positional pick-up and drop-off times and rerouting of some buses to make sure this is a safe alternative. (Legislative Council 2002b: 15)

There is a reference here to the 'measurement' of space, 'the primary schools are in fact only about 600 to 700 metres apart' and to the process by which these short distances are to be negotiated through transport solutions. Here, for the DET, space is relationally neutral, to be traversed, not to be engaged with. It is a pragmatic, technical construction that presupposes unchanging variables, and it may operate to close down arguments. It positions 'distance' as unproblematic, and constitutes it as a minor concern purged of hazards and free of contingencies.

Conversely, others note that space is constituted by interrelations. For example, Ms Stavert, a New South Wales Teachers Federation representative, stated that:

> I think it is absolutely clear that the people who originally made the decision about these communities had no knowledge of these communities, because on a map it all looks—Waterloo is up the road, Alexandria is over there, Erskineville is up there, but from my experience of working with people in these communities, they might as well be 20 kilometres away from each other because they are distinct communities. (Legislative Council 2002d: 18–19)

According to this discourse, space is about performance and is performative, students and parents are not just traversing 'empty space', they are actively engaged in constituting and being constituted by, spaces and places. The people who challenged the school closures mobilized a discourse of space that invoked the relational space/s of possibility that re-envisioned inner city schooling and the inner city. Underpinning the policy contestation was the 'battle for truth/s' about inner city schooling that mobilized 'legitimizing' and 'illegitimizing' space discourses for political purposes. The politics of place and space are evident in the testimony of Ms Munroe, chairperson of Redfern Primary School Parents and Citizens Association, who stated:

> Thank you for the opportunity to speak today. I would like to begin by acknowledging that the land we stand on today is Kuttabul land. I think by making that acknowledgement I come to you with a whole different aspect about history and learning and teaching, Aboriginal history, our sense of place. We have a very strong affinity with the Redfern community, we have a very long, proud association with the Redfern school....
>
> A sense of place: Alexandria certainly does not have it. Redfern Primary School has it, a sense of place, a sense of history, a sense of pride.... I can go to Redfern Primary School and point out the history of Gough Whitlam and Vincent Lingarie [and the struggle for land rights]. I cannot do that in any other school in that local area. I can go to that school and talk about whose land it is, and all the other pieces of Aboriginal land in the close vicinity to Redfern School.... That aspect of history is being forgotten in this whole process, our contribution of Aboriginal people to the Redfern community and to the Redfern area in particular, and our strong association with that school. (Legislative Council 2002d: 16–17)

In Ms Munroe's testimony the differentiated notion of relational space is evident, through the mobilizing of a set of spatial relations concerned with colonialism, racism, and dispossession, in a sense the postcolonial aspects of space (see Gregory 2004; Jacobs 1996; King 2003). Yet this notion of relational space appears tied to the 'illegitimacy' of Aboriginal attachments to place, an attachment that runs counter to the 'scientificity' of the DET's demographic arguments, and again Ms Munroe's mobilization highlights how the *attachments* of relational space contrast with the apparent *neutralities* of absolute space.

CONCLUSION

The empirical focus of this chapter was on the political mobilization of space discourses and education policy contestation in the inner city. I examined how certain spatial knowledges were mobilized as legitimate or illegitimate, particularly around notions of 'scientificity', during the 'battles over the status of truth' that were a part of the *Building the Future* policy initiative in inner Sydney. I was particularly interested in how 'scientificity' is able to be mobilized in tandem with 'absolute space' to make strong 'truth' claims about not only educational provision but also the constitution of the inner city. I think, at the least, this type of analytic can compellingly demonstrate how different types of discourses are mobilized and how certain types of 'space' (e.g. 'abstract') are seen as more legitimate than others (e.g. 'relational').

In this chapter, I was also interested in methodological and theoretical foci that were aimed at exploring the possibility of a poststructural spatial analytic in education policy sociology; the possibility of adding to the policy sociology 'analytical toolbox'. This possibility was explored by: one, looking at the usefulness of a particular takeup of poststructural criticism in education policy sociology, based predominantly on the work of Michel Foucault and Stephen Ball; and, two, exploring whether this form of analysis could be supplemented by drawing on notions of a poststructural geography. I see this work in a similar vein to the 'mapping [of] spaces and *re-presenting* them, [and] therefore...*de-naturaliz[ing]* them' (Gregory 1994: 277, original emphasis). In 're-presenting' the tacit assumptions of educational policy directions in the inner city, a poststructural spatial analytic assists in identifying different conceptions of space mobilized in educational policy change. Furthermore, a poststructuralist notion of space may constitute 'regimes of truth' as both historical and spatial and thus, hopefully, creates further possibilities for the 'sphere of the political' within education policy sociology.

ACKNOWLEDGMENTS

I am deeply grateful to Eva Bendix Petersen for her insistence that I write about poststructuralism and space with a sensitivity to not only argument, but also to form. All errors, both of content and style, are, of course, mine. I would also like to thank Colin Symes for his helpful comments.

NOTES

1. In the Australian Federal system of government the States have primary responsibility for the provision of education; however, the Federal Government can affect provision through funding mechanisms enabled through the Constitution (see Taylor et al. 1997).
2. Thus, for example, in this chapter I do not specifically take up questions of the subject which have been a significant aspect of poststructural thought (see St Pierre 2000: 500–505).
3. In this chapter I do not deal with the possibilities of scale (e.g. Collinge 2005; Robertson 2006).
4. However, it is also important to pay heed to Derek Gregory's point that 'thirdspaces' can also be, following Giorgio Agamben, 'spaces of the exception', such as refugee detention camps in Australia and Camp X-Ray at Guantánamo Bay in Cuba. As Gregory notes '[a]s these zones of indistinction multiply around the world, so it becomes clear that "thirdspaces"...are not

always and everywhere the emancipatory formations that some writers have taken them to be' (Gregory 2004: 258).

5. See Gregory (1994: Chapter 2) for how this played out in human geography.

REFERENCES

Agnew, J. (2005) 'Space: place', in P. Cloke and R. Johnston (eds.), *Spaces of geographical thought: deconstructing human geography's binaries*, London: Sage.

Ball, S. J. (1990) *Politics and policy making in education: explorations in policy sociology*, London: Routledge.

—— (1993) 'What is policy? Texts, trajectories and toolboxes', *Discourse: the Australian journal of educational studies*, 13.2: 10–17.

—— (1994a) *Education reform: a critical and post-structural approach*, Philadelphia: Open University Press.

—— (1994b) 'Some reflections on policy theory: a brief response to Hatcher and Troyna', *Journal of Education Policy*, 9.2: 171–182.

—— (2000) 'Performativities and fabrications in the education economy: towards the performative society', *Australian Educational Researcher*, 27.2: 1–23.

—— (2006) *Education policy and social class: the selected works of Stephen J. Ball*, London: Routledge.

Ball, S. J., Bowe, R. and Gewirtz, S. (1995) 'Circuits of schooling: a sociological exploration of parental choice in social class contexts', *Sociological Review*, 43: 52–78.

Ball, S. J., Maguire, M. and Macrae, S. (1998) '"Race", space and the further education market place', *Race, Ethnicity and Education*, 1.2: 171–189.

Ball, S. J., Maguire, M. and Macrae, S. (2000) 'Space, work and the "new urban economies"', *Journal of Youth Studies*, 3.3: 279–300.

Blackmore, J. (2002) 'Restructuring public schooling: a commentary on the New South Wales proposal Building the Future', *Journal of the HEIA*, 9.1: 37–46.

Bowe, R., Ball, S. J. and Gold, A. (1992) *Reforming education and changing schools: case studies in policy sociology*, London: Routledge.

Butler, J. (1997) *Excitable speech: a politics of the performative*, New York: Routledge.

Clegg, S. R. (1989) *Frameworks of power*, London: Sage.

Cloke, P. and Johnston, R. (2005) 'Deconstructing human geography's binaries', in P. Cloke and R. Johnston (eds.), *Spaces of geographical thought: deconstructing human geography's binaries*, London: Sage.

Collinge, C. (2005) 'The différance between society and space: nested scales and the returns of spatial fetishism', *Environment and Planning D: Society and Space*, 23.2: 189–206.

Crang, M. and Thrift, N. (2000) 'Introduction', in M. Crang and N. Thrift (eds.), *Thinking space*, London: Routledge.

Davies, B. (2004) 'Introduction: poststructuralist lines of flight in Australia', *International Journal of Qualitative Studies in Education*, 17.1: 3–9.

Doel, M. (1999) *Poststructuralist geographies: the diabolical art of spatial science*, Edinburgh: Edinburgh University Press.

—— (2000) 'Un-glunking geography: spatial science after Dr Suess and Gilles Deleuze', in M. Crang and N. Thrift (eds.), *Thinking space*, London: Routledge.

—— (2004a) 'Poststructuralist geographies: the essential selection', in P. Cloke, M. Crang and M. Goodwin (eds.), *Envisioning human geographies*, London: Arnold.

—— (2004b) 'Waiting for geography', *Environment and Planning A*, 36: 451–460.

—— (2005) 'Deconstruction and geography: settling the account', *Antipode*, 37.2: 246–249.

Foucault, M. (1972) *The archaeology of knowledge*, New York: Pantheon Books.

—— (1980) *Power/knowledge: selected interviews and other writings 1972–1977*, New York: Pantheon Books.

—— (1986) 'Of other spaces', *Diacritics*, 16: 22–27.

Gregory, D. (1994) *Geographical imaginations*, Oxford: Blackwell.

—— (2004) *The colonial present*, Oxford: Blackwell.

Gregson, N. and Rose, G. (2000) 'Taking Butler elsewhere: performativities, spatialities and subjectivities', *Environment and Planning D: Society and Space*, 18: 433–452.

Gulson, K. N. (2004) 'Education policy, urban renewal and identity: a spatial analysis of global change', Unpublished Ph.D. diss., Macquarie University, Sydney.

—— (2005) 'Renovating educational identities: policy, space and urban renewal', *Journal of Education Policy*, 20.2: 147–164.

—— (2006) 'A white veneer: education policy, space and "race" in the inner city', *Discourse: Studies in the Cultural Politics of Education*, 27.2: 251–266.

Howlett, M. and Ramesh, M. (1995) *Studying public policy: policy cycles and policy subsystems*, Oxford: Oxford University Press.

Jacobs, J. M. (1996) *Edge of empire: postcolonialism and the city*, London: Routledge.

Jensen, O. B. and Richardson, T. (2004) 'Framing mobility and identity: constructing transnational spatial policy discourses', in J. O. Bærenholdt and K. Simonsen (eds.), *Space odysseys: spatiality and social relations in the 21st century*, Aldershot: Aldgate.

Keith, M. (1997) 'Conclusion: a changing space and a time for change', in S. Pile and M. Keith (eds.), *Geographies of resistance*, London: Routledge.

Kenway, J. (1990) 'Education and the Right's discursive politics: private versus state schooling', in S. J. Ball (ed.), *Foucault and education: disciplines and knowledge*, London: Routledge.

King, A. D. (2003) 'Cultures and spaces of postcolonial knowledges', in K. Anderson, M. Domash, S. Pile and N. Thrift (eds.), *Handbook of cultural geography*, London: Sage.

Lefebvre, H. (1991) *The production of space*, Oxford: Blackwell.

Massey, D. (1993a) 'Politics and space/time', in M. Keith and S. Pile (eds), *Place and the politics of identity*, London: Routledge.

—— (1993b) 'Power-geometry and a progressive sense of place', in J. Bird, B. Curtis, T. Putnam, G. Robertson and L. Tickner (eds.), *Mapping the futures: local cultures, global change*, London: Routledge.

—— (1994) 'Double articulation: a place in the world', in A. Bammers (ed.), *Displacements: cultural identities in question*, Bloomington: Indiana University Press.

—— (2005) *For space*, London: Sage.

Murdock, J. (2006) *Post-structuralist geography: a guide to relational space*, London: Sage.

New South Wales Department of Education and Training (2001a) *Building the Future: an education plan for inner Sydney* (draft proposal), Sydney: New South Wales Department of Education and Training.

——— (2001b) *Building the Future: consultation report*, Sydney: New South Wales Department of Education and Training.

New South Wales Parliament Legislative Council: General Purpose Standing Committee No 1 (2002a) *Proposed closure and restructuring of inner Sydney schools*, Sydney: New South Wales Parliament Legislative Council: General Purpose Standing Committee No 1.

——— (2002b) *Report of 7th June 2002 proceedings before inquiry into the proposed closure and restructuring of government schools in inner Sydney*, Sydney: New South Wales Parliament Legislative Council.

——— (2002c) *Report of 29th May 2002 proceedings before inquiry into the proposed closure and restructuring of government schools in inner Sydney*, Sydney: New South Wales Parliament Legislative Council.

——— (2002d) *Report of 31 May 2002 proceedings before inquiry into the proposed closure and restructuring of government schools in inner Sydney*, Sydney: New South Wales Parliament Legislative Council.

Olssen, M. (2004) 'Neoliberalism, globalisation, democracy: challenges for education', *Globalisation, Societies and Education*, 2.2: 231–275.

Ozga, J. (1987) 'Studying education policy through the lives of policy makers', in S. Walker and L. Barton (eds.), *Changing policies, changing teachers*, Milton Keynes: Open University Press.

——— (1990) 'Policy research and policy theory: a comment on Fitz and Halpin', *Journal of Education Policy*, 5.4: 359–362.

Peters, M. A. and Burbules, N. C. (2004) *Poststructuralism and educational research*, Lanham,MD: Rowman and Littlefield.

Peters, M. A. and Humes, W. (2003) Editorial: 'The reception of post-structuralism in educational research and policy', *Journal of Education Policy*, 18.2: 109–113.

Petersen, E. B. (2004) 'Academic Boundary Work: The discursive constitution of 'scientificity' amongst researchers within the social sciences and humanities', Ph.D. diss. No.25, University of Copenhagen, Denmark.

Richardson, T. and Jensen, O. B. (2003) 'Linking discourse and space: towards a cultural sociology of space in analysing spatial policy discourses', *Urban Studies*, 40.1: 7–22.

Robertson, S. L. (2002) 'The politics of re-territorialisation: space, scale and teachers as a professional class', *Curriculo sem Fronteires*, 2.2: xvii–xxxiv.

Robertson, S. L. (2006) 'Absences and imaginings: the production of knowledge on globalisation and education', *Globalisation, Societies and Education*, 4.2: 303–318.

Smith, N. (2004) 'Space and substance in geography', in P. Cloke, M. Crang and M. Goodwin (eds.), *Envisioning human geographies*, London: Arnold.

Smith, N. and Katz, C. (1993) 'Grounding metaphor: towards a spatialized politics', in M. Keith and S. Pile (eds.), *Place and the politics of identity*, London: Routledge.

Soja, E. W. (1989) *Postmodern geographies: the reassertion of space in critical social theory*, New York: Verso.

——— (1996) *Thirdspace: journeys to Los Angeles and other real-and-imagined places*, Cambridge, MA: Blackwell.

St Pierre, E. A. (2000) 'Poststructural feminism in education: an overview', *International Journal of Qualitative Studies in Education*, 13.5: 477–515.

Stronach, I. and MacLure, M. (1997) *Educational research undone: the postmodern embrace*, Buckingham: Open University Press.

Strüver, A. (2004) '"Space oddity": a thought experiment on European cross-border mobility', in J. O. Bærenholdt and K. Simonsen (eds.), *Space odysseys: spatiality and social relations in the 21st century*, Aldershot: Aldgate.

Taylor, C. (2001) 'Hierarchies and "local" markets: the geography of the "lived" market place in secondary education provision', *Journal of Education Policy*, 16.3:197–214.

Taylor, S., Rizvi, F., Lingard, B. and Henry, M. (1997) *Educational policy and the politics of change*, London: Routledge.

Youdell, D. (2004) 'Engineering school markets, constituting schools and subjectivating students: the bureaucratic, institutional and classroom dimensions of educational triage', *Journal of Education Policy*, 19.4: 407–431.

4 Space, equity, and rural education

A 'trialectical' account

Bill Green and Will Letts

> [T]he way we imagine space has effects....
>
> (Massey 2005: 4)

Space is commonly seen as an essentially neutral category, and moreover as 'completely transparent, unmediated and therefore utterly unknowable' (Rose 1993: 70). Educational space is typically seen therefore as a 'container' within which education simply 'takes (its) place', with varying degrees of effectiveness and efficiency. From classrooms and schools to larger administrative jurisdictions such as 'districts', 'regions', and 'clusters'—all such sites and territories operate as commonsense, convenient registers of educational organization and governance, activity and application. Following a long tradition of programmatic scepticism, however, the question can and should be asked: Are they? What if space is *problematized*, in an explicit Foucaultian sense? What would result from according spatial relations and dynamics a more deliberate, focused attention, with space taken seriously as a matter of interest and concern, for policy and pedagogy alike? What happens when space is thought differently, and *other* spaces are drawn into calculation?

This chapter takes questions such as these as both its incitement and its topic, in exploring the relationship between what might be called spatial difference-dynamics and educational politics in the field of rural education. Our work has focused on the notions of space, equity, and rurality, and links with other investigations of space, communication, and power in curriculum and schooling. This present account of the conjunction of education, geography, and governmentality is situated within a distinctive educational–administrative jurisdiction, namely New South Wales, Australia, and it is accordingly informed by, among other things, a larger historical geography of Australian education. Our concern here is how best to understand and represent educational space, with its characteristic configurations and challenges of power, difference, identity, and disadvantage.

A note on 'method': we offer here what, following Soja (1996) can be called a *trialectical* account of rurality and education, space and equity. Such an account seeks to bring together 'historicality', 'sociality', and 'spatiality', again in Soja's (1996: 72) terms, or the intellectual perspectives and resources of the historical, the social, *and* the spatial imaginations. But it is also, and importantly, a certain practice of writing and thinking. In this latter sense, the chapter is therefore something of an *experiment*—but one that is consistent, we argue, with various lines of flight in contemporary theory, including geography, cultural studies, and philosophy.[1] We are concerned with practical problems of representation and rhetoric, then, as well as what we are coming to see as the necessarily transdisciplinary and even transgressive character of the matter at issue here, in this larger field of enquiry and debate. In what follows, we build a case for thinking differently about issues of space, equity, and rural education, and contextualize these discussions with issues arising out of our work in rural (teacher) education.

ON SPACE AND EDUCATION

Space as such would appear one of the most under-examined concepts in educational theory and practice, as indeed it is in other fields of inquiry and endeavour. 'The lack of a theoretical understanding of "space" within mainstream sociology', Stehlik (2001: 34) writes, 'means that it is "missing" within the development of sociological research inquiry.[... R]arely, if ever, do we identify space as a variable for analysis'. For Usher (2002), linking new awareness of spatial relations and dynamics to the so-called modernism–postmodernism debate, means that:

> Space is now more and more seen as having been under-theorised and marginalised in relation to the modernist emphasis on time and history. This emphasis constructed space as neutral, fixed and immobile, unrelated to the social and without impact on the formation of subject identity and biography. (Usher 2002: 41)

He goes on, however, to acknowledge and to argue that what is really at issue is a proper sense of the articulation of time and space, a new sensitivity to the significance of the hyphen ('space-time'): 'we need to think now in terms of "space-time", of a conception and actuality of time and space as inseparable and interactively relational' (Usher 2002: 42).

This understanding is developed most substantively in Doreen Massey's (1993, 1994) work on the politics of 'space-time' and the concept of 'power-geometry', and the relationship between 'the spatial organization of society...and the production of the social' (Stehlik 2001: 34). Harvey (1989, 1993, 1996) is similarly a key reference, exploring notions of geography, new and enduring formations of class, postmodernity, technocultural change, social justice, and what he perhaps notoriously termed 'time-space compression'. Geographers such as Soja (1989, 1996) and D.M. Smith

(1994), and also Rose (1993), are also important sources for rethinking space and society, and informing educational research accordingly. This relatively recent link with educational research is especially appropriate and timely, given Marsden's (1977: 21) observation, almost three decades ago now, that '[e]ducational phenomena are distributed in space'—something which now might indeed almost be called a key defining element, however naturalized, in the practice and institution of education.

Foucault is another particularly useful source in linking space to considerations of power and knowledge in a manner that greatly illuminates curriculum history and rural education alike. As he asserted in one of his interviews, '[s]pace is fundamental in any form of communal life; space is fundamental in any exercise of power' (Foucault 1984: 252). This particular interview is extremely interesting and particularly pertinent here, in fact, especially when read in relation to his work on governmentality. He ranges across architecture and the problem of urban space, observing in so doing that the model of the city served from the seventeenth century on as an organizing principle for 'the government rationality that was to apply to the whole territory' (Foucault 1984: 241): 'A state will be well organized when a system of policing as tight and efficient as that of the cities extends over the entire territory'. He later goes on to note the emergence of 'three great variables—territory, communication, and speed' (Foucault 1984: 244).

A ready link exists here back to the historical account of curriculum and schooling in Australia (Seddon 2001; Green 2003a). The point to focus on is that an argument might well be made, accordingly, that public education systems as they consolidated in the mid- to late-nineteenth century were conceived and organized on the basis of what can be called a *metro-centric* model, or the norm of the modern city. Inasmuch as the new educational bureaucracies were concerned with the management of schools and schooling, and hence populations, they were equally concerned with the management of territories. This model applied first to the more immediate space of the city, and then to the outward movement of settlement. How might financial and human resources, which are always limited, be used to best effect, and due accountability and control ensured, in organizing and overseeing educational space?

Such an account has obvious implications for thinking about rural education. Could it be that the historical provision of a network of rural schools across Australia, together with an elaborate system of inspection and eventually its organization in terms of administrative territories, is best understood as the application of a panoptic grid, a spatializing logic of 'super-vision' and surveillance? If that is indeed the case, then the traditional concern of state bureaucracies for uniformity of provision across their jurisdictions, linked as that always has been to particular notions of equity, becomes more intelligible. At the same time, it becomes possible to think differently about this, and to draw in more explicitly Foucaultian notions of knowledge, power, and space, as a new framework for understanding rural education, in both policy and practice.

SITUATING RURAL EDUCATION

It is hard to contemplate rural education in Australia, as perhaps elsewhere in the developed world, without becoming 'historical'. Rural schools seem to be deeply inscribed in the national imaginary, as is the bush myth more generally (Green 2003b)—illustrated quite dramatically, and projected to a global audience, in the opening and closing ceremonies of the Sydney Olympics. Indeed there is much that can be said about the enduring contribution and significance of the small bush school in Australia's social and educational history, and in its development as a complex democratic society and a distinctive modern nation. Yet Australia is now very clearly a highly urbanized country, with the bulk of its population living in the capital cities hugging the coast. Overall, the country is characterized in fact by sprawling high-density metropolitan patterns of both residence and work, and a declining rural-regional sector. This is notwithstanding the fact that rural-regional Australia continues to contribute significantly to the national economy. Yet there is no doubting the long-term effect of globalization and neoliberalism in this regard (Gray and Lawrence 2001; Lawrence, Gray and Stehlik 1999), or the continuing significance of Australia, as uniquely a 'nation-continent', being relatively speaking sparsely populated and also spread across a huge landmass.

This is directly connected to the importance of geography in Australian education, and also its taken-for-grantedness. Already located in the Antipodes, on the other side of the globe, the original set of colonies was later to become six distinct states and territories. Seddon (2001) among others has pointed to the characteristic standardization and uniformity of Australian curriculum and schooling, right from the outset, with the emergence and consolidation of highly centralized and bureaucratized public education systems. As she writes: '[f]rom the late 1880s to the 1970s, State Education Departments exercised virtually unconstrained authority with State-systems of public education', something that was 'substantially realized through the State curriculum...oriented to the building of each State' (Seddon 2001: 315). She explicitly points to the 'connection between geography and sovereignty, which is central to colonisation' (Seddon 2001: 311). Green similarly observes that the challenging task confronting the new educational bureaucracies was one of managing territories as well as populations, and 'mapping them onto each other' (Green 2003a: 136). This involves developing an analysis that brings together 'governmentality, gender and geography' as indeed 'a key distinguishing feature of Australian education viewed historically, especially when its vast, sparsely populated spaces and distances are taken into full account' (Green 2003a: 136). A feature of the public system as it consolidated was its *centralization*, with schools and schooling administered from a head office located in each of the capital cities. An early review of Australian education made the following assertion:

> The truth is that centralization is the best form of educational rule for a young country with a vast hinterland. It ensures that the children of the rural pioneer receive as good an education as the children of the banker or artisan in the city, and Australia is proud of the efficiency of her small country schools. (Browne 1927: xvii–xviii)

Other commentators note the manner in which standardization and simultaneity have long gone hand-in-hand in Australian education, with locational differences and disadvantages thereby more or less annulled (Cole 1932: x–xi; Seddon 2001: 315). What is interesting here is the view, deeply enshrined in the Australian system, that location is of no consequence to the delivery of education, that distance can be effectively annulled, and that space simply doesn't matter—at the same time as there is clearly a concern for adequacy of coverage, equity, and efficient forms of management and surveillance.

The point of making these historical observations is that Australian education arguably remains geography-blind today, now as much as ever—notwithstanding what might be called the operational importance of geography. In one project we have been engaged in, addressed to public educational provision in inland NSW, an immediately striking feature was the huge size of the rural education endeavour, taking in as it does approximately 84 percent of the State's landmass and yet just 13 percent of its total population. A key organizing principle for the project was that *place matters* (McConaghy and Burnett 2004), and now that the project is effectively finished, we are confirmed in the view that proper and informed contextualization is critical in the reconceptualization of both teacher education and school pedagogy, especially in and for rural conditions and circumstances. Although rurality is not to be defined or delimited by geography, let alone determined by it, nonetheless geography is clearly an important consideration. This means among other things taking into account matters of distance and terrain, as well as location, or what might be better described as locational relativity, all of which have implications for and effects on educational access and equity. Yet while there would appear to be widespread acceptance of the significance of place, the same cannot be said, even now, about its corollary, space. Indeed, policy would seem space-blind. This raises for us some concern about the apparent acceptance of the notion of place too, since it seems highly likely that the two are intricated and interrelated conceptually.[2] Nonetheless, place seems somehow more congenial to the policy imagination, in this circumstance at least. For the purposes of this chapter, then, we want to pose the following question: If place matters, what about space?

SPACE, POWER, AND THE RURAL

Of course there is an important sense in which thinking about the *rural* in this context is already a matter of spatial differentiation—a 'dividing practice'—in that it is customarily and often set against the *urban*, as categories of classification. However, this is a limited distinction ('rural-urban'), as Hugo (2002) among others argues (e.g. Moriarty, Danaher, and Danaher 2003), needing itself to be supplemented by a further distinction, between what Hugo calls accessibility and isolation, or remoteness. This is partly because some ostensibly rural locations are clearly urban centres, i.e. small towns and regional cities. The distinction, however, is better understood as referring to the socio-spatial relationship between the capital cities in each state, as a greater metropolitan area, and the rest of the country, located progressively more inland from the coastal fringe. Furthermore, the important point to be made here is that what is most significant in this regard is that some areas are more privileged and powerful than others, and hence what must be accounted for are what Thomson (2002) describes as 'geographies of distinction', and indeed of inclusion and exclusion. Living in the larger cities is, for many, a more attractive prospect than living in the country, generally speaking, although further internal discriminations certainly need to be made here, regarding both the 'city' and the 'country'.

That spatial relations are implicated in the production of subjectivity might well be explored through the notion of a distinctive 'Australianness', or of the (stereo)typical Australian character, and relatedly through Australian inflections of postcolonial identity. The very notion of the 'Antipodes' is relevant here, as marking an Other place, the other side of the world, and of (Western) civilization as it was known then. So too is the iconic image-set of the 'beach' and the 'bush', each a distinctive figural location and a key element in the national imaginary: symbolic places of fantasy, fear, and desire as much as work and play. The so-called bush myth, evoked earlier, is part of this whole complex formation, long enshrined in popular culture and poetry lessons, although significantly of urban origin, as Davison (1994) has argued. A similar case can be made of the notion of 'countrymindedness', originally introduced in Australian political science and cultural history and later 'widely adopted as a label for the discourse that constitute (*sic*) the Australian rural "world view"' (Aitkin 1985; Share 1995: 10). An always contradictory and ultimately unsatisfactory concept, it can be seen nonetheless as an indicator of the relationship between space and subjectivity, and of the manner in which knowledge, power, and space come together. The point to emphasize here is the significance of *culture*, and of cultural production, in re-thinking space and identity.

At this point we want to look more closely at the concept of rurality itself. This is clearly a complex and much debated concept, and one that seemingly eludes ready understanding or simple definition. Nonetheless, as various commentators observe, rurality must be understood socially, and as a social construction; moreover, as 'a multidimensional concept, symbolically constructed and both historically and cultural contingent, which is difficult to define and measure' (Bourke and Lockie 2001: 9). In commonsense terms and common usage, it seems to bring together 'connections to agriculture, wide-open spaces, isolation and self-reliance' (Bourke and Lockie 2001: 10). Our particular focus here is on space and rurality. For the Swedish sociologist Goran Therborn, '[r]urality, the defining criterion of rural sociology, is, of course, a spatial concept' (Therborn 1999: 21). As he writes: 'Spatial orders and relationships are extremely important to social organization' (Therborn 1999: 22). His account takes in notions of modernity and globalization, seeing in social change and development a significant spatial dimension, moreover one that is a 'still largely under-theorized and underdeveloped aspect of social analysis' (Therborn 1999: 33). He sees in rurality, and rural sociology, an opportunity for sociology more generally to strengthen its insights into the social, accordingly, with a particular role to play in understanding changing formations of global modernity.

His attention to the spatial is echoed in accounts such as those of Halfacree (1993) and Lobao (1996), from within rural sociology, who share with him and other commentators a sense of rurality's complexity as well as its marginality. Lobao (1996: 77) sees a significant 'spatial legacy' in rural sociology, with 'space' providing 'a sense of subject matter and boundary' for the field. Space and rurality have been at the core, from the outset. Importantly she sees such notions as marginalized in sociology more generally, which has tended to be organized by particular constructions and priorities of modernity (Lobao, 1996: 79). In this context, she argues for 'a sociology of the periphery', which she glosses as motivated by an interest in 'those settings where the attributes of capitalism of the core play out unevenly'—where 'the people and places of concern remain those marginalized in the course of development' (Lobao 2001: 80). For Halfacree, rurality has been conceptualized as a category characterized either by 'spatial determinism' or 'spatial indifference', neither of which is adequate. Rather, what is needed is a critical socio-spatial framework, a social-dynamic theory of space and spatiality whereby there is a dialectical relationship between space and society, practice and representation, and an emphasis on the social production of space. In this, his account is congruent with those of Lefebvre (1991) and Soja (1989, 1996), among others. Halfacree is especially concerned with what he describes as 'the rural as social representation' (p. 29), and hence the shift to a postmodern understanding of space and rurality, and a movement 'away from a concentration on tangible space in favour of the non-tangible space of "social representations"' (p. 23). That is something increasingly evident, in fact: a growing emphasis in such accounts on changing conceptions and constructions of space within postmodernity.

Soja's work becomes particularly relevant in this regard. In arguing for the reassertion of space in critical social theory, he works from the following assertion:

> We must be insistently aware of how space can be made to hide consequences from us, how relations of power and discipline are inscribed into the apparently innocent spatiality of social life, how human geographies become filled with power and ideology. (Soja 1989: 6)

Far from being neutral, or natural, he seeks to problematize all our common usages and understandings of the spatial in social life and everyday discourse, as a field of reference. His objective, as he puts it, is:

> to encourage [us] to think differently about the meanings and significance of space and those related concepts that compose and comprise the inherent spatiality of human life: place, location, locality, landscape, environment, home, city, region, territory, and geography.(Soja 1996: 1)

Drawing on Foucault and Lefebvre, Soja locates space firmly within the field of power and the social. Spatiality takes on analytic and conceptual significance alongside and interwoven with 'historicality' and 'sociality', as organizing categories. A key notion he develops is that of 'real-and-imagined places', drawn from his account of what he calls 'Thirdspace', which he glosses as a rich view of social space 'that builds on a Firstspace perspective that is focused on the "real" material world and a Secondspace perspective that interprets this reality through "imagined" representations of spatiality' (Soja 1996: 6). Soja's focus is on Los Angeles as an emblematic 'real-and-imagined' place and on cities more generally. However, it is our contention that the rural is very generatively seen in these terms, and in particular that rural Australia is usefully understood this way, as a 'real-and-imagined' place par excellence. This has various implications and challenges for rural education, as we shall go on to argue. For the moment, however, we want to stay with Soja's distinction, following Lefebvre, between what he identifies as *perceived* space, *conceived* space, and *lived* space, or between the 'physical', the 'mental', and the 'social', as three distinct but related registers of human spatiality. Paulston and Liebman (1996: 25) gloss this as 'a new unitary theory of space' within which:

> space is simultaneously a spatial practice (or externalised, material environment), a representation of space (a conceptual model used to direct practice) and a space of representation (the lived social relation of users to the environment).

What such a formulation enables is a far richer view of space than that to be observed either in commonsense or policy, one that brings together geographical and environmental features and processes, calculations and programmes, or 'designs', and human practices, the actual and the imaginary. Crucial to this emerging framework is a focus on *power*. This is partly in relation to knowledge and discourse, as in Foucault, but also from a neo-Marxist and postcolonialist point of view, with reference to ruling relations, exploitation, and social control. Space is never innocent in such an account, therefore. What happens, happens somewhere, and on a number of levels, or 'scales'. Taking due account of this situated complexity of social practice becomes a matter of principle and urgency. We turn now to consider more directly a particular instance of rural education in the light of the socio-spatial framework outlined here.

RURAL SCHOOLING IN NEW SOUTH WALES

New South Wales (NSW) is one of seven states and territories making up the contemporary Australian nation. With clearly the largest population overall, it is also among the largest geographically, area-wise. However, the bulk of the population live in a megapolis comprising Newcastle, Sydney, and Wollongong, with the remainder spread out across the state, beyond the Great Dividing Range and up and down the coastal fringe. Sydney itself is acknowledged to be a truly global city, with all that entails, in terms of economic development, social complexity, and environmental effect. A legacy of Federation, NSW shares with the other Australian states and territories an overriding administrative responsibility for its own educational jurisdiction—a vast territory, in fact. The public system, historically by far the largest, is organized by what is currently called the Department of Education and Training (DET), a large educational bureaucracy, whose Head Office is located in central Sydney. Traditionally, education in NSW has been highly centralized. While this is a feature of the Australian system more generally, it has been arguably most strongly articulated in NSW.

The relationship between space and equity is directly implicated here. The historical argument for centralized organization has always been linked to uniformity of provision, whether in terms of the location and availability of schools, the provision of staffing, or the question of resources. Writing in the mid-1920s, in the first comprehensive overview of education in Australia, Browne (1927: xvii) argues that there are very good reasons for the 'Australian States possess[ing] the most strongly centralized systems of educational administration in the world'. As he writes: 'It should be remembered that Australia's educational problem is unlike that of any other nation. Here is a whole continent as large as the United States and nearly as large as Europe'. Moreover, '[a]s yet it contains only six million people' (Browne 1927: xviii).[3] Population and territory come together in a problem *par excellence* of educational administration. Given the spread of a relatively small population over vast territories, how do the pertinent authorities manage the task of providing adequate forms of schooling—moreover fairly? The answer seemed to lie in highly regulated practices and centrally mandated syllabi, and overall a system of surveillance and control. This latter, as in other countries (e.g. Canada—see Curtis 1988), was realized principally by way of an inspectorate. From the mid-nineteenth century on:

> [t]he inspector was the most important, and often the only, link between the educational administration and each of its schools.[...] By the early 1850s,... both New South Wales and Tasmania were beginning to embrace the ideal of a professional, full-time inspectorate and a decade later it was an integral part of both state education systems. (B. Smith 1990: 77)

Importantly, each inspector was assigned a particular responsibility, a distinctive area, often literally a territory: a specific educational space. Hence arose the practice of organizing education into specific jurisdictions, or 'districts', the effect being to produce a grid of classification and accountability, a cellular complex, a panopticon. Moreover, each inspector was responsible for preparing a report which then became incorporated into a dense network of disciplinary writing, an 'archive'. It is important to see this not in terms of imposition, however, nor as simply the bearing down of (monolithic) authority; rather, as an exercise in liberal government, whereby the task is one of taking due care of schooling and its various constituencies. Care in this case goes hand-in-hand with control. For our purposes here, the partitioning of educational-administrative space in this way is to be seen as a way of providing for equity, in the sense that it allows for a dual logic of care and control to be realized, both locally and as it were globally.

From the outset, a constant concern for state governments and education departments alike was with the provision of adequate and appropriate forms of schooling for rural areas and communities. The population rapidly spread out across the continent, and each of the states and territories found themselves organized in terms of a major city and its hinterland, with distinctive patterns of settlement emerging in accordance with the prevailing environmental and geographical conditions. The bush school became a powerful motif in both policy and the popular imagination (Nelson 1989). Providing such schools, in all sorts of locations, became the particular responsibility and challenge for public education. This is something now enshrined in national expectations, in the Adelaide Declaration's assertion that schooling should be 'socially-just,' with students' outcomes unconstrained by 'differences arising from students' socio-economic background *or geographic location*' (MCEETYA 1999; our emphasis). In NSW, this principle has been whole-heartedly endorsed from the outset—hence initiatives such as the 'travelling schools' and the correspondence school (Freeman 1993) of the early twentieth century, and the establishment of the NSW's second Teachers College at Armidale in 1928. The practice however has been mixed, and arguably continues to be so.

A DIVIDING SPATIAL PRACTICE?

In 2002 we started researching, along with colleagues, the relationship between rural schooling and teacher education in NSW. A collaborative undertaking between two rural-regional universities and the DET, the project took as its focus what were at the time 11 school districts, all located in the inland area of NSW and all 'west of the divide'—as noted previously, approximately 84 percent of the State's landmass. This involved, again approximately, 13 percent of the State's total population, and over 500 of the State's public schools. The State's public education system was itself organized in terms of 40 school districts overall, which meant that the project was addressed to just over 25 percent of these jurisdictions. The majority of these jurisdictions, just fewer than 75 percent, were located therefore in just 16 percent of the State. Each of these school districts was overseen and managed by a superintendent, thereby continuing a longstanding administrative tradition in Australian education.

From its inception, the project was interested in matters of space and locality. Among other things, the research team was motivated to 'thinking anew about issues of space and place, distance and isolation'. To that end, we saw one of our key aims as 'producing a social and educational cartography: a mapping of educational provision, opportunity and disadvantage in rural NSW'. The project undertook 'a spatial-analytic investigation of the distinctive character of rural and remote education' (Green et al. 2001: 4), with specific reference to NSW. Along with this explicit conceptual focus on space and spatiality, there were therefore two already spatialized starting points for the research in question: first, the decision to work from the admittedly anecdotal reference to the 'disadvantage' said to be experienced by those located west of the Great Dividing Range—on the other side of the mountains; and second, the happy coincidence that this area corresponded to the way the public education system was organized at the time—that is, into the aforementioned 11 school districts. In retrospect, this was a good example of what Bourdieu (1992) describes as the 'pre-constructed', and something to be wary of, accordingly (see also Bourdieu and Wacquant 1992: 229); although it must be said, too, that it was arguably a factor in making the project intelligible to our 'industry partners', in that it coincided with what may have been their own preconceptions, prejudices, and investments, as well as our own. The point to make here is that these are already spatial-geographic distinctions.

A further key feature of the project was its focus on social justice, and on the forms and degrees of social and educational disadvantage widely supposed to be associated with life in rural and remote settings. The project was in part motivated by a perceived sense of the 'unevenness' of educational outcomes in the country, compared with State averages and norms—that is to say, 'variable learning outcomes for children in [rural] New South Wales, as a result of their socio-economic positioning and geographical location' (Green et al. 2001: 2). Was there a demonstrable rural disadvantage? In this regard, we came to see our concern as being a question of the geographical over-determination of social and educational disadvantage. To that end, we wanted to produce what we called a social map of educational disadvantage in inland NSW, and perhaps a social atlas of rural education more generally—although it might fairly be said that this could be readily subsumed in a larger cartography of public education (see Teese 2000). We quickly realized that this was far too ambitious an undertaking, on the one hand, and on the other, far too political, especially given we were working with the DET as our industry partner. Nonetheless, the project's somewhat attenuated explorations of educational achievement data in the study area, although currently under embargo at the time of writing, are congruent with our initial hypothesis. Moreover, with regard to the teaching turnover in the inland areas, that is, staffing stability and 'staffing churn' (Thomson 2002: 112), there is a clear spatial pattern emerging, a sharp distinction between what we identified as an 'inner ring' of mainly younger, less experienced teachers and an 'outer ring' of more established teachers (Green and Novak 2006). This is a matter of some real concern, especially given the growing research consensus that quality teachers and teaching genuinely matter to educational success. Moreover, other investigations conducted almost contemporaneously into rural-remote patterns of school staffing (Roberts 2004) and the nature and provision of public education in NSW (Vinson 2002) provide substantial support for the account emerging in our study, and we feel confident accordingly that it is valid.

That said, we want to focus now more specifically on the issue of space and equity. This seems a dramatically under-theorized issue not just in rural education but in education more generally, including education policy. What needs also to be realized is that there is, to say the least, an under-appreciation of how space is implicated in social and educational practice, at every level— that space matters, in short, and how space matters. Rather than being neutral or natural, as much education policy seems to understand it:

> the spatial is socially constituted. 'Space' is created out of the vast intricacies, the incredible complexities, of the interlocking and the non-

> interlocking, and the networks of relations at every scale from local to global. (Massey 1994: 265)

This is, in part, a matter of thinking differently and otherwise about space, about spatial relations and spatial practices, of problematizing space, and taking account of constructs like 'Thirdspace' as a reference-point for policy *and* practice. In this latter regard, as Danaher, Danaher, and Moriarty (2003: 166) assert, 'any account of contemporary rural education needs to attend to *both* the materiality and the mentality of living and learning in rural locations'—a point we thoroughly endorse, and resume in a moment. More importantly here, there is an urgent need to de-naturalize notions of distance, isolation and location, and indeed geography itself, as much more than simply a landscape, an accident of history and circumstances. Like space, geography itself is not innocent, but must be seen as both a resource for, and a medium of power. As Smith writes:

> Geographical space is clearly implicated in social exclusion. Social interaction usually implies physical propinquity (modern means of communicating over long distances notwithstanding), and to separate people spatially can be an effective means of exclusion as well as control. (D.M. Smith 1994: 45)

Although this appears to fly in the face of those sometimes breathless accounts that extol the virtues of the new technology in overcoming distance, taking a historical and comparative view of such a view is likely to be sobering, and unsettling. Which groups are located where, and why, and what are the effects of this?

MAPPING DISADVANTAGE

As a final engagement with our theme, then, we want to look more closely at the case of 'Dubbo', one of our original school districts. It no longer exists as such, or rather it has been subsumed into a new regional re-organization of public education in NSW—a different kind of spatial framing, operating at a larger scale.[4] 'Dubbo' was clearly one of the largest spatial units in the previous system, ranging from just short of the central west to the state border. More than two thirds of the district rated as remote or very remote on the Accessibility/Remoteness Index of Australia (ARIA), and nowhere in the district was rated as highly accessible (Commonwealth Department of Health and Aged Care 2001). The following, taken from the District Office's 2002 Annual Report, indicates its scope:

> Measuring the distance from Dubbo District Office (DDO) to each of the district's schools and campuses makes this the largest educational

> district in NSW. One indicator of distance is that from January-November 2002, the car fleet used by DDO staff registered one million kilometres. This is equivalent to one and a half return trips to the moon.

Other factors such as climate and terrain clearly need to be factored in here. At the time, the district was responsible for 61 schools and campuses, including nine high schools, 14 central schools, 12 large primary schools, and 21 small primary schools. A feature of the district was its Aboriginal enrolment—the highest proportion of the 40 school districts across the State.[5] It is fair to say that test results and other educational indicators across the district were (allegedly) such as to represent a real challenge, both politically as well as from the more specific point of view of policy and practice.

The official map of the district indicated very clearly a definite pattern of centre–periphery dynamics, locationally at least, with the district organized around the large town of Dubbo and administered from there. Another lesser pattern could be discerned in relation to the smaller town of Bourke, located on the Darling River. The project sought to overlay two further maps on this, one depicting social disadvantage (Vinson 2004) and the other, the distribution of the Aboriginal population, including its patterns of concentration.[6] Although by no means uniform, there was a marked geography of distinction emerging here: an articulation of socio-economic disadvantage and indigenousness, of race and class. This became all the more marked the further west one looked, again with some obvious counter-indicative pockets, or 'patches'. Less tangible, and clearly significant, was another feature of the social and educational cartography at issue here: the movement of Aboriginal people up and down the Darling, a pattern of mobility and transience, with some Aboriginal students proving understandably difficult to pin down, especially for the purposes of testing and accountability. Space in this instance is clearly 'multidimensional, fluid, and mobile' (Danaher, Danaher, and Moriarty 2003: 165). Moreover, even when we are looking at a space as seemingly homogenized as a single school district, or even a single town within that district, Massey's (1993: 67) assertion that 'clearly places do not have single, unique "identities"; they are full of internal differences and conflicts' is certainly borne out by the evidence in this instance.

With the administrative centre of the State's educational system being Sydney, 'Dubbo' was perhaps understandably positioned as being 'out there' (Stehlik 2001). Such a characterization links to and draws on enduring cultural scripts about 'centre-periphery' dynamics, and a poignant history of racism and the effective erasure of Aboriginal presence. With respect to envisioning wide open spaces ('out there') on maps of various kinds, Piper (2002: 9) writes:

> sovereignty…became linked to erasure, based on the notion of creating a territorial blank slate on which one could construct colonial rule and authority. Clearing space, in effect, became a way to establish whiteness—or to differentiate oneself from [I]ndigenous peoples.

From some vantage-points, therefore, although the entire district seemed to be 'out there,' within 'Dubbo' itself, as already noted, Dubbo city was seen as the centre, the 'big smoke', with the larger field of the periphery stretching up to the far north-western reaches of the district.

In terms of equity, thinking spatially about communities allows us to, as Stehlik (2001: 37) writes, 'consider issues of power and identity; incorporate difference; challenge centre/periphery arguments; incorporate history and geography in our analyses; and recognize alternative forms of knowledge'. Instead of thinking of spaces as neutral (if, indeed, we think of them at all), thinking about equity spatially entails making explicit the relations of power and knowledge often tacitly obscured in traditional ways of thinking about spaces and the maps that purport to represent them. Instead, spaces are imbued with histories that render them politically, economically and socially charged. We must think of spaces literally, that is, but also as knowledge spaces, as comprising both 'materiality' and 'mentality'. In this regard, as Gray and Lawrence (2001: 95) note, 'regional Australia suffers in particular from the historical constraint of the colonial legacy and a lack of resources in its relationship with metropolitan Australia'. One very noticeable legacy of colonialism is that racial tension still seethes in many parts of the country—meaning here both Australia in its entirety, and also the country as in 'out there'. Referring specifically to racially-charged fantasies and anxieties in the far west of NSW, Morris (2001: 244) notes the following:

> The asymmetrical power relations are writ large in the prevailing conditions of social existence, health and life expectancy and a highly disproportionate rate of criminalisation, which clearly demarcate the social and material conditions of the powerful and the powerless.

In this way, the 'west' embodies a 'contact zone' (Pratt 1992: 4), consisting of 'social spaces where disparate cultures meet, clash, grapple with each other often in highly asymmetrical relations of domination and subordination'. In this regard, as Morris (2001: 246) writes: 'such contested terrain is not spatially determined by physical geography, but created as part of the racial, historical and cultural geography of the modern settler state formation'. The lack of knowledge about and understanding of rural and remote lives and places therefore 'simply represents a continuation of the constant colonial struggle to fill voids in maps' (Piper 2002: 6). But merely thinking of this space in terms of what the maps reveal keeps hidden the ways that space is an arena for cultures to interact—to collide, to enmesh, to resist, to coexist. This occludes issues related to power, identity, and discourse, relegating them to the periphery of attention as irrelevant, as invisible. Especially when unproblematized, maps can hide the ways that this particular space 'is an heterogeneous space where there is a contest over the meanings of racism, of equality, of history and colonialism and its aftermath' (Morris 2001: 247). Thinking about 'Dubbo' is, then, an exemplary instance of the notion of 'real-and-imagined places', in Soja's resonant formulation. Already consigned to history, as it were, it nonetheless maintains an awkward presence; it lingers still, in memory, in records and conversations, even as it is sustained in the here-and-now of praxis, in the ongoing work of rural schooling and the western region. A fuller analysis of 'Dubbo', and indeed of inland NSW, would require appropriate means of display of the available visual data, as well as more opportunity to tease out patterns and features of pertinence and note. Such work needs still to be done. In this case, it might well join up with, and supplement, other work on the geography of racism in NSW (Dunn and McDonald 2001). Educationally, such an account seems to us to be clear indication of the relationship between, and significance of, space and equity in rural education.

CONCLUSION

In this chapter, we have been concerned to demonstrate the manner in which space matters in and for education, and hence for education policy. We have sought to do so partly by working with the notion of space itself as a problem of representation. Something of an *experiment*, this has meant deliberately mixing the conceptual and the empirical, 'mentality' and 'materiality', the Real and the Imaginary. This has particular resonance for thinking critically about rural education, which has increasingly been marginalized in mainstream sociology of education, as elsewhere, with the city coming to be seen as not simply the unspoken norm of progressive educational thought but also the organizing centre of educational reform, in more ways than one. Modernity arguably has privileged time over space, as various commentators have argued. As a distinctive field of inquiry, education studies has similarly and perhaps accordingly been dominated by notions of time and change, of progress and reform, with movement from the countryside to the city being seen as a marker of the inevitability of social development. In this story of progress and development, space has been largely unproblematic, at best a technical issue of communication and control, policy and practice. Yet it is that very obviousness, or rather its *obliviousness*, that emerges now as an issue of theoretical and political import, as space takes on a significance that it has previously been denied, perhaps especially in education. The rural is arguably a socio-spatial phenomenon *par excellence*, and something deeply inscribed in the history and discourse of Australian education. How then to re-think and re-imagine the social and geographic differences that matter now, in a new era of globalization, environmental degradation, and geo-political tension?

Historically, the equity project of public schooling involved seeking to treat not simply every child but also every place as the same, and accordingly to work towards equitable provision of educational resources and services. Space was effectively neutralized, or denied—or rather, every attempt was made to overcome the so-called 'tyranny of distance' and the vagaries of terrain. At the same time, heroic efforts were made, practically, to work within the context, and sometimes against the grain, of an always challenging, chastening geography. Now, in postmodern times, it may be that space is (re-)asserting itself, as above all else a field of differences, and of the differential production of identity and value. Within this, the rural might well be understood not just as a matter of geographical difference, but as increasingly the site of the Other, real-and-imagined, within a spatial economy predicated increasingly on a differential distribution of power and privilege. Space hides consequences from us all, as Soja (1989: 6) has suggested. In this instance, thinking differently about what happens 'out there', at a distance, is a matter of both urgency and obligation.

NOTES

1. Not the least of which is recent work on space and spatiality (e.g. Massey 2005; Crang and Thrift 2000).
2. While sometimes these terms ('place', 'space') seem to be used interchangeably, and elsewhere we look more specifically at the notion of *place*, drawing on work such as that of Cresswell (2004) and others, our larger concern is with *space*, which in fact might well be regarded as a superordinate category. On some of the problematical ways in which place is typically understood, see Massey (2005: 5–6) for a particularly succinct statement.
3. The sentence in fact continues thus: '...but these people are almost entirely of British stock', and then: 'Can [Australia] keep her continent white?'—a reminder of the enduring politics of space *and* race in Australia's history. We resume this theme later in this chapter.
4. The shift has been from 'districts' to the larger unit of the 'region', within which are 'School Education Areas' (subsequently renamed as 'School Education Groups')—with little indication of a theorized view of space or geography behind such policy decisions. 'Dubbo' as such is now part of a much larger 'Western Region'.
5. In this regard, it is worth noting the following: 'Aboriginal students continue to be the most educationally disadvantaged student group in Australia' (NSW AECG and NSW DET 2004: 12).
6. Here, we would like to acknowledge the work of Arthur Read and, in particular, Steven Gibbs, of the Spatial Data Analysis Network (SPAN) at Charles Sturt University, Bathurst. The project compiled an archive of several hundred maps, of various kinds, influenced as it was by the notion of *social cartography* (Paulston and Liebman 1996), and this is still awaiting full analysis.

REFERENCES

Aitkin, D. (1985) 'Countrymindedness: the spread of an idea', *Australian Cultural History*, 4: 34–41.

Bourdieu, P. (1992) 'Thinking about limits', *Theory, Culture and Society*, 9: 37–49.

Bourdieu, P. and Wacquant, L. J. D. (1992) *An invitation to reflexive sociology*, Chicago: University of Chicago Press.

Bourke, L. and Lockie, S. (2001) 'Rural Australia: an introduction', in S. Lockie and L. Bourke (ed.), *Rurality bites: the social and environmental transformation of rural Australia*, Annandale, NSW: Pluto Press.

Browne, G. S. (1927) 'Introduction', in G. S. Browne (ed.), *Education in Australia: a comparative study of the educational systems in the six Australian States*, London: Macmillan.

Callejo Perez, D. M. and Swidler, S. A. (2004) 'Re-imagining rural schools as historical and spatial places for research and reform', Paper presented at the Annual Meeting of the American Educational Research Association, San Diego, USA, April 12–16.

Cole, P. R. (1932) 'Introduction', in P. R. Cole (ed.), *The primary school curriculum in Australia*, Melbourne: Melbourne University Press.

Commonwealth Department of Health and Aged Care (2001) *Measuring remoteness: Accessibility/remoteness index of Australia*, Canberra: Department of Health and Aged Care. Accessed at: http://www.health.gov.au/ari/aria

Cresswell, T. (2004) *Place: a short introduction*, Oxford: Blackwell.

Curtis, B. (1988) *Building the educational state: Canada West, 1831–1871*, Philadelphia: Falmer Press.

Crang, M. and N. Thrift. (eds.) (2000) *Thinking space*, London: Routledge.

Danaher, P. A., Danaher, G. and Moriarty, B. (2003) 'Space invaders and pedagogical innovators: regional educational understandings from Australian occupational travellers', *Journal of Research in Rural Education*, 18.3: 164–169.

Davison, G. (1994) 'Sydney or the Bush: an urban context for the Australian legend', in P. Russell and R. White (eds.), *Pastiche 1: reflections on nineteenth century Australia*, Sydney: Allen and Unwin.

Dunn, K. M. and McDonald, A. (2001) 'The geography of racism in NSW: a theoretical exploration and some preliminary findings from the mid-1990s', *Australian Geographer*, 32.1: 29–44.

Foucault, M. (1984) 'Space, knowledge and power', in P. Rabinow (ed.), *The Foucault Reader*, Harmondsworth: Penguin.

Freeman, A. (1993) 'The travelling schools of New South Wales 1908–1949', *Education in Rural Australia*, 3.1: 7–18.

Gray, I. and Lawrence, G. (2001) *A future for regional Australia: escaping global misfortune*, Cambridge: Cambridge University Press.

Green, B. (2003a) 'Curriculum inquiry in Australia: towards a local genealogy of the curriculum field', in W. F. Pinar (ed.), *International handbook of curriculum research*, Mahwah: NJ: Lawrence Erlbaum.

—— (2003b) 'Curriculum, public education and the national imaginary: re-schooling Australia?' in A. Reid and P. Thomson (eds.), *Towards a public curriculum*, Canberra: Australian Curriculum Studies Association.

Green, B., McCulla, N., Boylan, C., McConaghy, C., Maxwell, T., Wallace, A., Meyenn, R., Chandler, W. and Brock, P. (2001) 'Productive partnerships for teaching quality: quality improvement, school-community practice and teacher education in and for rural and remote settings', *Project Outline: ARC Discovery Application*.

Green, B. and Novak, M. (2006) 'Staffing and the provision of quality teaching in inland rural New South Wales', in B. Green (ed.), *Spaces and places: the NSW rural teacher education project—a report*, Bathurst: Charles Sturt University.

Halfacree, K. H. (1993) 'Locality and social representation: space, discourse and alternative definitions of the rural', *Journal of Rural Studies*, 9.1: 23–37.

Harvey, D. (1989) *The condition of postmodernity*, Oxford: Basil Blackwell.

—— (1993) 'From space to place and back again: reflections on the condition of postmodernity', in J. Bird, B. Curtis, T. Putnam, G. Robertson and L. Tickner (eds.), *Mapping the futures: local culture, global change*, London: Routledge.

—— (1996) *Justice, nature and the geography of difference*, Oxford: Blackwell.

Hugo, G. (2002) 'Australia's changing non-metropolitan population', in D. Wilkinson and I. Blue (eds.), *The new rural health*, South Melbourne: Oxford University Press.

Lawrence, G., Gray, I. and Stehlik, D. (1999) 'Changing spaces: the effects of macro-social forces on regional Australia', in C. Kasimis and A. Papadopoulos (eds.), *Local responses to global integration*, Aldershot: Ashgate.

Lefebvre, H. (1991) *The social production of space*, Oxford: Basil Blackwell.

Lobao, L. (1996) 'A sociology of the periphery versus a peripheral sociology: rural sociology and the dimension of space', *Rural Sociology*, 61.1: 77–102.

Marsden, W. E. (1977) 'Historical geography and the history of education', *History of Education*, 6.1: 21–42.

Massey, D. (1993) 'Power-geometry and a progressive sense of place', in J. Bird, B. Curtis, T. Putnam, G. Robertson and L. Tickner (eds.), *Mapping the futures: local culture, global change*, London and New York: Routledge.

—— (1994) *Space, place and gender*, Cambridge: Polity Press.

—— (2005) *For space*, London: Sage.

McConaghy, C. and Burnett, G. (2004) 'Education in the country: what's the difference?' *Prime Focus: Journal of Australian Primary School Leaders*, 36: 28–29.

Moriarty, B., Danaher, P. A. and Danaher, G. (2003) 'Situating and interrogating contemporary Australian rural educational research', *Journal of Research in Rural Education*, 18.3: 133–138.

Morris, B. (2001) 'Policing racial fantasy in the far west of New South Wales', *Oceania*, 31.3.

MCEETYA (1999) *The Adelaide declaration on national goals for schooling in the twenty-first century*, Canberra: Ministerial Council on Education, Employment, Training and Youth Affairs.

Nelson, H. (1989) *With its hat about its ears: recollections of the bush school*, Crows Nest, NSW: ABC Enterprises.

NSW Aboriginal Education Consultative Group and NSW Department of Education and Training (2004) *The Report of the review of Aboriginal education*, Sydney: NSW Department of Education and Training.

Paulston, R. G. and Liebman, M. (1996) 'Social cartography: a new metaphor/tool for comparative studies', in R. G. Paulston (ed.), *Social cartography: mapping ways of seeing social and educational change*, New York: Garland.

Piper, K. (2002) *Cartographic fictions: maps, race, and identity*, New Brunswick, NJ: Rutgers University Press.

Pratt, M. (1992) *Imperial eyes: travel writing and transculturation*, London: Routledge.

Roberts, P. (2004) *Staffing the empty schoolhouse: attracting and retaining teachers in rural, remote and isolated communities*, Sydney: NSW Teachers Federation.

Rose, G. (1993) 'Some notes towards thinking about the spaces of the future', in J. Bird, B. Curtis, T. Putnam, G. Robertson and L. Tickner (eds.), *Mapping the futures: local culture, global change*, London: Routledge.

Seddon, T. (2001) 'National curriculum in Australia? A matter of politics, powerful knowledge and the regulation of learning', *Pedagogy, Culture and Society*, 9.1: 307–331.

Share, P. (1995) 'Beyond "countrymindedness": representation in the post rural era', in P. Share (ed.), *Communication and culture in rural areas*, Wagga Wagga: Centre for Rural Social Research, Charles Sturt University.

Smith, B. (1990) 'William Wilkins' saddle bags: state education and local control', in M. R. Theobald and R. J. W. Selleck (eds.), *Family, school and state in Australian history*, Sydney: Allen and Unwin.

Smith, D. M. (1994) *Geography and social justice*, Cambridge: Basil Blackwell.

Soja, E. W. (1989) *Postmodern geographies: the reassertion of space in critical social theory*, London: Verso.

—— (1996) *Thirdspace: journeys to Los Angeles and real-and-imagined places*, Oxford: Blackwell.

Stehlik, D. (2001) '"Out there": spaces, places and border crossings', in S. Lockie and L, Bourke (eds.), *Rurality bites: the social and environmental transformation of rural Australia*, Annandale, NSW: Pluto Press.

Teese, R. (2000) *Academic success and social power: examinations and inequality*, Melbourne: University of Melbourne Press.

Therborn, G. (1999) 'Modernity, globalization and the rural world', in C. Kasimis and A. Papadopoulos (eds.), *Local responses to global integration*, Aldershot: Ashgate.

Thomson, P. (2002) *Schooling the rustbelt kids: making the difference in changing times*, Crows Nest, NSW: Allen and Unwin.

Usher, R. (2002) 'Putting space back on the map: globalisation, place and identity', *Educational Philosophy and Theory*, 34.1: 41–55.

Vinson, T. (2002) *Pubic education in NSW: report of the 'Vinson Inquiry'* Annandale, NSW: Pluto Press/ NSW Teachers Federation.

—— (2004) *Community adversity and resilience: the distribution of social disadvantage in Victoria and NSW and the mediating role of social cohesion*, Richmond, NSW: Jesuit Social Services.

5 Geographical information systems (GIS) and school choice

The use of spatial research tools in studying educational policy

Chris Taylor

Geographies of education are not new. One of the earliest attempts to highlight the spatial imbalance in education provision across the UK was by Henry Brougham during the 1820s, using a series of national surveys. Furthermore, as Bradford (1990) has outlined, it is not difficult to see the role a geography of education can have in understanding society (e.g. through housing markets), the economy (e.g. through labour markets), and polity (e.g. through political power). There have been numerous social studies during the last few decades where space and place have played an important part in understanding social inequalities and the inadequacy of policy in addressing these, particularly in relation to the notion of 'community' and 'neighbourhood'. For example, in their study of working class children Jackson and Marsden (1966) place the community of Huddersfield in the UK at the centre of their analysis and understanding of the impact of education. Yet it is still the case that the geography of education is a relatively underdeveloped area of research (Bradford 1990; Taylor 2002). This is beginning to change, and this book is testament to that. To a large extent work in this area has been primarily interested in more contemporary sociological theories of space of, say, David Harvey and Doreen Massey. But there is much also to be gained by considering more traditional theories of space, such as accessibility, location, and mobility. It is also the case that there have been significant technological innovations in geographical analyses through the use of Geographical Information Systems (GIS) that are entirely appropriate in their application to the field of education. This chapter endeavours to illustrate how more traditional theories of space, alongside some of the more contemporary spatial theories, can aid our understanding of school choice and the impact of increased marketization in publicly-funded education provision. It also provides an illustration of the use of GIS in this area and in particular the use of postcode data analysis that is now commonplace throughout marketing research and 'quantitative' geography.

SCHOOL CHOICE IN THE UK

The UK has seen an acceleration of open enrolment in compulsory schooling during the 1980s and 1990s through successive Conservative and Labour governments. The 1980 Education Act, the 1988 Education Reform Act, and the 1997 Admissions Code of Practice have notably shifted the responsibility for allocating school places from local education authorities to parents and schools. Open enrolment, or more precisely the opportunity for parents to express a preference for the school of their choice—otherwise known as 'parental choice'—has been used by both Conservative and Labour governments as an important vehicle for school reforms, primarily in raising educational standards and examination performance. Similar developments have occurred throughout Europe, Australia, New Zealand, and the United States, alongside other market-driven reforms in the delivery of education and public services generally. Open enrolment has been used alongside the local management of schools, per pupil funding, programmes of school diversity, and the publication of school examination league tables to create a quasi-market in state schooling that is driven by the parameters of economic supply and demand. 'Successful' schools will benefit from increased popularity, while 'less successful' schools will need to address their unpopularity by being more responsive to the needs of local parents and the demands of government (see Taylor 2002; Gorard et al. 2003).

It is important to note, however, that school choice is not an entirely new phenomenon in UK education. Some parents and families have always had a choice between and within the state and private education sectors. Some local authorities have always given parents the option of applying for a place in a school of their choice (Dore and Flowerdew 1981). It is also necessary to distinguish between the freedom of school choice, as tested more recently in the law courts, and the practice of school choice as determined primarily by the admissions arrangements employed by local authorities and individual schools. Although there have been changes to the dominant form of admissions arrangements in the UK, most notably from academic selection (during the tripartite system of provision) to catchment areas (during the comprehensive era of schooling), and more recently open enrolment (alongside the new public sector management in education and other public services), it is still the case that there is great diversity (temporal and geographical) in how admissions are managed and how school places are finally allocated once parents have stated their preferences (White et al. 2001).

This history and diversity of school admissions forces us to re-examine the notion of school choice in three ways. The first is that school choice does not only occur in the annual ritual of school admissions. The consumption of education can be more complex and can occur in a number of ways and at different times. For example, choice of school is an important aspect of residential choice, and the housing market is inextricably linked to the education quasi-market. Therefore, much school choice, with or without open enrolment (i.e. at the time of the admissions process) has already occurred when choosing a place to live.

The second key issue with regard to school choice is the distinction between the process of school choice and the outcomes of school choice. The use of the concepts of social and cultural capital (Bourdieu and Passeron 1977) in school choice research gives primacy to 'agency', and hence many studies explore the processes of school choice and how parents from different socio-economic backgrounds have responded to and engaged with the education quasi-market. However, the apparent neglect of 'structure' in these studies has meant that a mismatch exists between the inequities found in the process of school choice and what can be considered the outcomes, or products, of school choice in terms of quality of provision, access to school places, and the social class composition of school intakes. Both the process and outcomes of school choice are important, and their distinction continues to frame the debate about the impact of open enrolment in the UK.

The third is that the choice of school is not space-neutral. When deciding what school their child should attend, parents are making complex decisions that are not the same as choosing a loaf of bread or health insurance. They are making decisions about mobility, which can be bounded by issues of accessibility. How easy is it to get to the schools? How long will it take to get there? Does travelling to that school fit in with other daily routines, such as going to work or doing the shopping? But these decisions are also framed in what Edward Soja (1980) terms the socio-spatial dialectic; the interconnected relationship between society and space/place. This, in turn, led Giddens (1984) to develop his structuration theory, drawing on processes of time-space routinization and time-space distanciation. So the choice of school can be determined by the routes that children (and parents) will take and how they feel about the neighbourhoods or communities that they must travel through in order to get to the school of their choice. Their choices can, then, be affected by both familiarity and fear, and managed by the availability of particular transport, such as the car or the public bus. The mobility of some families in a locale can be greater than others, and hence the accessibility of different schools in both physical and emotional terms can be greater for some than others. Furthermore, the notion of motility (Kaufmann 2002), or mobility capital, can be seen as distinct to economic, social, and cultural capital that are often used to help explain school choices (see Ball et al. 1995).

The purpose of this chapter is to argue that, to help resolve these three tensions, a 'geography of school choice' is required. Not only does this help us to consider alternative ways in which inequities may have occurred, such as through the housing market, admissions rules, or over time, but it also helps highlight other significant aspects of school choice that tend to be overlooked, such as the way in which families from more working class backgrounds engage in school choice and are 'consuming' education.

TOWARDS A GEOGRAPHY OF SCHOOL CHOICE

A geographical perspective of school choice is not new, and is not confined to geography. Four early examples of this are: the work of Michael Adler and colleagues in Scotland (1989), which examined the early stages of open enrolment to primary and secondary schools in Scotland; a study by Ron Glatter and colleagues as part of their PASCI (Parents and School Choice Initiative) survey, a relatively large study on school choice around the Midlands; research by Stephen Ball and colleagues that followed a number of families in detail through the process of transferring from primary to secondary school; and the work of Diane Reay, which has explored the meaning and sensitivities of school choice from the varying perspectives of 'others,' including the children involved and working class families (Reay and Lucey 2000). These studies have, in turn contributed towards a geography of school choice in differing but important ways. For example, Adler et al. (1989) were the first researchers to compare and contrast the movement of pupils between catchment areas, highlighting an uneven geography of choice across two cities. Simultaneously, Glatter (1989) was developing a spatial model of school choice based on proximity. Ball et al. (1995) emphasized the social geography of school choice with their focus on the sociological landscapes of education. And Reay and Lucey (2000), for example, considered children's geographies in the contribution to, and understanding of, school choice.

Each of these studies contributes to an overall understanding of two important qualities of a geographical perspective on school choice. The first is the 'lived' contexts in which school choice occurs. This could be the way in which the school choice by individual families is constrained by the location and distance of possible schools, the way in which different families perceive the availability of schools and school places, and the relationship between school reputations and the social and residential geography in which they are located. One of the most important features from such an approach to the study of school choice is the notion of local education markets and competition spaces (Taylor 2002). The second quality of a geographical perspective is the natural, often spatial, comparators that emerge. Not only do these allow us to identify and highlight differences (and similarities) in the processes and outcomes of school choice, they also help identify possible explanations or causal links between certain phenomena. These intrinsic comparators are essential if we are to really understand, for example, the impact of open enrolment on education and society.

An important development in spatial theory that may help understand how families from different socio-economic backgrounds engage with the education market and school choice is in the notion of motility. 'Motility can be defined as the capacity of entities (e.g. goods, information or persons) to be mobile in social and geographic space, or as the way in which entities access and appropriate the capacity for socio-spatial mobility according to their circumstances' (Kaufmann et al. 2004: 750). Such an approach would be interested in the relationship parents and their children have with mobility, and what competences, capabilities, and skills they have in providing the potential for mobility, and therefore school choice. Clearly there will be some relationship between motility and economic and social wealth, e.g. in the access to private car transportation. But such an approach would also consider the familiarity and fear of getting to schools that may cross socio-economic cleavages. To what extent such theoretical developments will improve our understanding remains untested, but it should be apparent that examining the spatial relationships and characteristics of school choice and the education market is required before fully appreciating the impact of such educational reforms.

GEOGRAPHICAL INFORMATION SYSTEMS (GIS)

One of the most significant technological developments for spatial analyses in recent years has been Geographical Information Systems (GIS). The definition of a GIS tends to depend on its particular application and who is developing it. However, there are three key components to a GIS (Heywood et al. 1998): that it is a computer system, incorporating hardware and software; that it uses spatially referenced data; and that it provides management and analytical capabilities. In effect it is a form of database tool that implicitly references the data over the Earth's surface. Furthermore, just like other technological developments used in social science research, such as statistical software and computer-aided qualitative data analysis software, the operator or user of the GIS is critical in ensuring it is used appropriately and that its capabilities are enhanced. Generic GIS software is readily available from major suppliers such as ESRI and MapInfo. Even Microsoft is now incorporating GIS technology into its Office software.

One of the reasons why GIS is being increasingly used in social science research is due to parallel developments in the role, collection, and availability of postcode data (Martin 1996). Postcodes have provided an easy and efficient source of geographical information. In the UK the government-funded Office for National Statistics (ONS) produces the All Fields Postcode Directory (AFPD). This contains all the postcodes used in the UK and identifies the administrative, health, electoral, and other geographies in which they are located. This has meant it is relatively straightforward to link individual households, via their postcodes, to other geographical information. It is also possible to spatially reference these postcodes for analytical purposes; for example, in measuring their proximity to key service providers. Postcodes can be spatially referenced in two ways: First, as a geographic *area*. The majority of postcodes represent an area that includes a particular number of households, approximately 15 households per unit postcode (Raper et al. 1992). Some postcodes represent a single building, such as a high storey apartment block, hospital, or prison, but remain as congruent geographic areas, albeit small areas. It is also possible to spatially reference a postcode as a point in space. This is useful in representing individual households in a GIS. The AFPD provides the grid reference coordinates of the bottom-left corner of the 100-metre grid containing the first address (alphanumerically) in the postcode. Although the resulting spatial coordinates can only be a proxy for the exact locations of households within a given postcode this can still be of enormous benefit in any spatial analysis of postcode data. Although measuring the accuracy of these spatial coordinates is difficult there have been numerous attempts to examine their accuracy. For example, Wilson and Elliot (1987) evaluated the 1986 version of the directory and found that 72 percent of postcodes were accurate to within 100 metres of their actual location, 93 percent were accurate to within 400 metres and 95 percent were accurate to within 900 metres. Clearly the geographical accuracy of households to postcode coordinates is greater in urban areas than it is in more dispersed rural areas. Although the proxy location for postcode areas can be problematic, it does, however, help preserve anonymity for individual households. It should also be noted that UK postcodes and the AFPD are updated every three months, so matching the appropriate postcode directory to the date when the postcode data were gathered or obtained improves their spatial accuracy.

The benefits of using GIS in social science research are not just confined to the use of postcode data. There are other important uses of GIS that are based on other sources of data. This can include Global Positioning Systems (GPS) to generate very accurate spatial coordinates, existing road and transport networks, and a wide variety of area-based data, such as spatially aggregated secondary data and school catchment or administrative areas.

GIS AND SCHOOL CHOICE

Research into school choice and open enrolment provides a very good example of the analytical capabilities of using GIS as a tool to research the impacts of educational policy. Early examples of its use in the context of the education market include Higgs et al. (1997) and Clarke and Langley (1996). The appropriateness of using GIS in this field is largely because school choice and school admission is a form of mobility. A child has to travel from their home to school and back nearly every day for up to 13 years, and this travel is all but compulsory (see McMillan 2005). These journeys constitute a significant amount of mobility in urban and rural areas. School choice is, therefore, spatially constructed and determined (among other factors). And since school choice is central to the marketization of compulsory education and schooling it lends itself very well to using spatial research tools in educational policy. This movement or mobility of pupils to schools can be analysed in spatial terms using the functionalities of GIS.

There are three main ways that GIS has been used to contribute to an understanding of the impact of increased marketization in education, and school choice in particular. Each of these will be discussed in turn, using examples from my own research into school choice and its impacts on school intakes over the last ten years. Much of this research has been relatively straightforward, using readily available spatial data and generic GIS software and functionalities.

The first analytic capability for a GIS in school choice research is its mapping tools. Although primarily descriptive and illustrative this is a useful vehicle for a consideration and appreciation of the varied contexts to the education market place. The simple production of Choropleth maps that describe the nature of school provision are important in understanding the variety of educational producers in the education marketplace. This mapping exercise can also highlight the importance of administrative boundaries in the education landscape, delineating the opportunities and constraints for school choice across space. An example of how this has been used in school choice research is illustrated by the existence of private-state continua in secondary education provision across England. Drawing upon Bradford's (1993) concept of the private-state continuum, six different local forms of private-state continuum can be identified based on the different combinations of Community, Grant Maintained (now Foundation Schools), Voluntary Aided and Voluntary Controlled, or Independent Schools in each Local Education Authority (Taylor 2001a). The resulting 'map' of school diversity across the UK illustrates the degree and form of educational privatization, and provides a key component of the context in which school choice operates (Figure 5.1).

There are other, additional analytical powers in the construction and presentation of such mapping using GIS. In particular it assists in examining the spatial correlation of factors or variables, i.e. the existence of two or more features across space. This is relatively straightforward using statistical techniques when the two (or more) datasets share the same spatial characteristics. But when looking at, for example, the relationship between school intakes (represented as points) and residential characteristics (represented by areas) this correlation requires additional geographical analytical tools. There are spatial tools in a GIS to calculate these relationships but mapping these variables as layers that overlap one another is equally as useful, just as plotting two variables against one another on a chart is useful. This method has been very important in identifying the relationship between the socio-economic mix of school intakes and the socio-economic mix of residential areas. In effect this has shown how levels of between-school

Figure 5.1 Distribution of local private–state continua in England.

segregation can be very closely related to differing levels of residential differentiation in some, but not all, areas of England and Wales. Figure 5.2 provides an example of this for one local authority in outer London. The schools are represented in their localities by circles—the bigger the circles the greater the proportion of their pupils eligible for free school meals. Eligibility for free school meals is predominantly determined by levels of household incomes. The underlying residential differentiation in Figure 5.2 is based upon levels of socio-economic deprivation of the household population, aggregated to the level of Enumeration Districts (ED) (typically 200 households per ED), living in the area. The Townsend Index of Deprivation is a composite measure based upon four key variables of the proportion of households: with no access to a car; that are overcrowded (more than one person per room); in owner-occupied accommodation; that contain unemployed economically active members of the household. The darker the shading the more socio-economically deprived the households are in each ED. Figure 5.2 clearly illustrates the relationship between the proportion of a school's intake that are eligible for free school meals and neighbour-

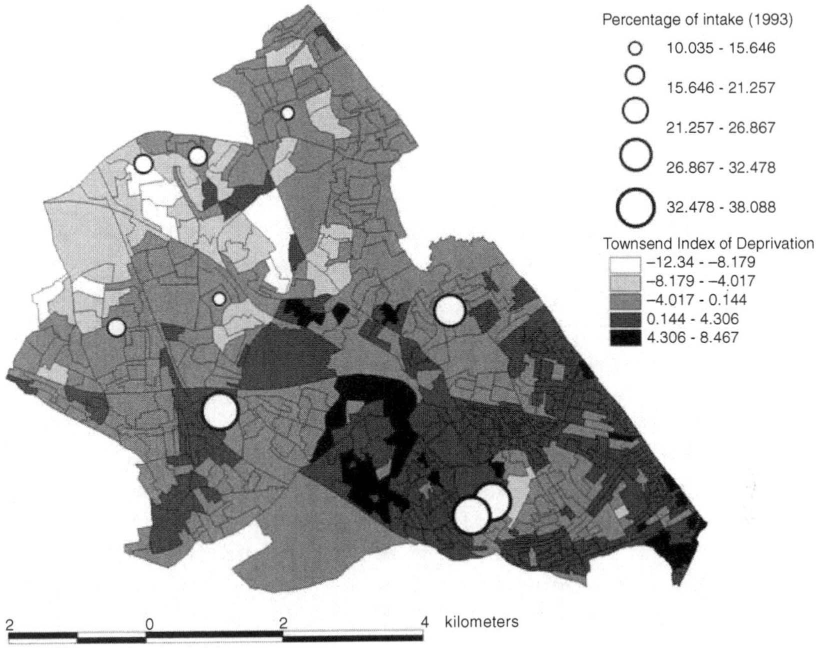

Figure 5.2 Proportion of school intakes, Outer London Borough.

ing levels of residential poverty—the greater the levels of residential socio-economic disadvantage in one area of the local authority the greater the proportion of the local school's pupil intake that are eligible for free school meals. However there are in contrast, other local authorities where the relationship between school intakes and residential characteristics are not as obvious (see Taylor and Gorard 2001a). The distinction between different local education markets is important, then, in appreciating the impact of school choice on between-school segregation.

The second major analytical capability of using GIS in the study of school choice is using the home postcodes of the children who attend different schools. Much school choice research is primarily concerned with the different socio-economic characteristics of the children who choose or attend different schools. From such research it is possible to gauge the different spatial features of that process and outcome of choice. But to systematically analyse the home locations of many school intakes would be extremely time-consuming and limited to those families participating in the study. Using the home postcodes of every child who attends all schools in a given area provides an exhaustive view of the spatial dimensions to school choice.

In the examples that follow, the children's postcodes were provided by different local education authorities attending all schools in their authority. Obtaining such postcode data can be problematic. For a start there may be some discrepancies in the precise location of the postcodes' spatial coordinates and the actual home location of each child, as discussed earlier. It is also the case that the home postcode for every pupil is not always recorded by the schools and the local authority accurately. Typically, when analyzing many postcode datasets from different areas of England and Wales postcodes are shown to be valid for approximately 95 percent of the pupil population. Checking that the missing children are distributed equally among all schools ensures that their absence from any analysis is significantly reduced. The coverage and accuracy of postcode data is improving, and with the collection of individual pupil data for every maintained school in England and Wales through the Pupil Level Annual Schools Census (PLASC), access to such data is now much greater. However, the greatest limitation to this dataset is that there are no records for pupils attending the independent schools sector or other non-maintained schools. Where such schools exist in an area or local education market specific permission to access their children's home postcodes must be gained separately. It is also important to note that the local education market may cross local authority administrative boundaries and hence, for completeness, data from several congruent authorities may be required.

There are numerous ways that the postcode data can be used in a geographical analysis of school choice and competition. The most obvious way this data can be used is to present and examine a real coverage of school intakes. Figure 5.3 illustrates the spatial coverage of just two school intakes in one local authority in England. The use of these two school intakes highlights the significant difference in the spatial dimensions of school choice. In particular, the distance and proximity of children to the schools they attend, and the homogeneity or diversity of communities and neighbourhoods where a school's intake comes from are highlighted.

Another example of the use of postcode data is to consider the gains and losses of children between every school in each local education market. Not only does this illustrate the apparent hierarchical nature of school choice to emerge across a given local education market (Taylor 2001b), it can also be used to highlight the uneven distribution of 'competition' between schools for pupils. In order to identify this competition it is necessary to identify what would be the traditional or local school for a child to attend. This is a theoretical decision, but one that is aided by the capabilities of a GIS. For example, the traditional or local school could perhaps be determined by which school 'catchment' area a child lives in—a typical model for allocating school places in the UK. These catchment areas could be digitized or drawn as an electronic coverage and entered into the GIS for further interrogation. Alternatively, it is increasingly more common in England and Wales for school places to be allocated according to distance or proximity between the school and the home of the child where schools are oversubscribed. In other words, the traditional or local school would be the one that the child lives nearest to. Although not described as catchment

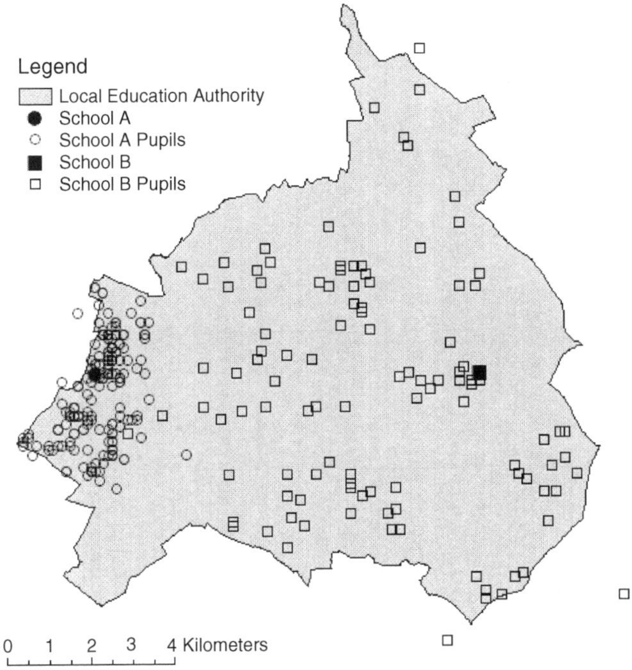

Figure 5.3 Spatial coverage of two secondary school intakes.

areas as such, this method of allocating school places can be presented in the form of areas. The GIS can be used to easily create Thiessen polygons, otherwise known as Voronoi or Proximal polygons, around each school. The result is that any location within these polygons or areas would always be closer to the respective school than any other school. Consequently, they can be used to identify which school every pupil lives nearest. Figure 5.4 illustrates an example of the 'surrogate' catchment areas around every secondary school in a local authority based on the proximity model of allocating school places. It is also the case, therefore, that manipulation of the polygons or catchment areas provides the potential to predict the impact on school intakes as a result of school reorganization or change in admission policy. For example, the impact of re-drawing catchment areas, amalgamating or closing schools, or changing the criteria in allocating school places can be projected in order to provide more confident predictions about the impact on school rolls and the composition of school intakes.

Using either method to define the local school for every child, it is then possible to identify the number of children who live in the local catchment area but who actually attend a different school. This can be calculated for every school and every child, resulting in a matrix of gains and losses between all schools in a given area. Figure 5.5, then, illustrates the resulting net gains (of five pupils or more) between schools in one urban

Figure 5.4 Thiessen polygon 'catchment' areas.

local education market. The direction of the arrows shows the direction of the net gain, and the thickness of the arrow indicates the significance of that net gain in terms of pupil numbers. In this example, the schools are located approximately to one another to preserve some anonymity of the particular urban education market represented. Not only does this show the complexity of school choice and competition in a given area (and note this only illustrates net gains, not *all* gains between schools), it also illustrates the hierarchical nature of school choice—note that School L largely gains pupils from other schools in the area, in contrast to School A that predominantly loses pupils to other schools—and the significance of particular 'competing' schools, such as the competition between schools J and M, between schools E and F, and between schools that are not necessarily adjacent to one another, such as between schools C and H, and between schools A and F. On further analysis of school choice in this local education market we see that these patterns are framed by the choices of particular social class and ethnic groups and by the different types of schools that are located in the area.

A more advanced use of GIS in analyzing patterns of choice, based on the home postcodes of individual children, is based on the use of gravity

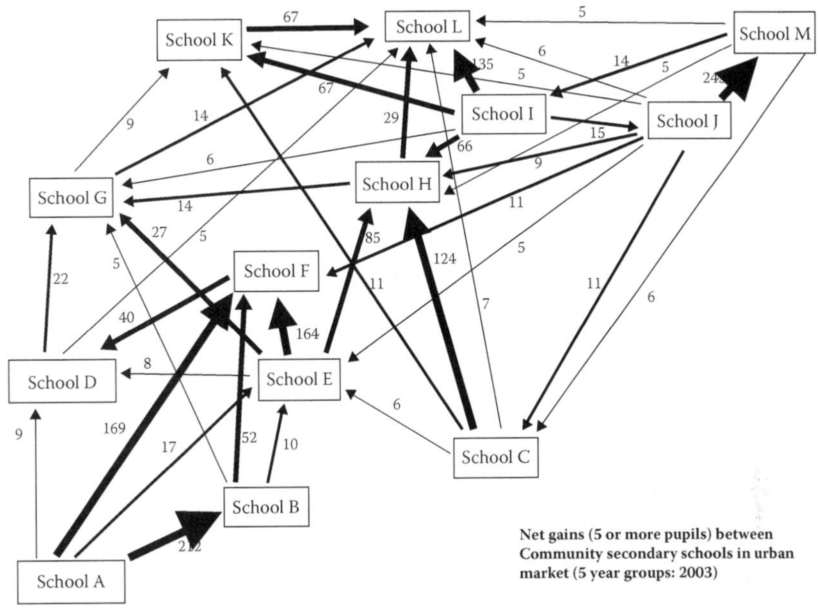

Figure 5.5 Net gains between secondary schools.

models. Using the integrated functions of a GIS it is possible to calculate 'probabilities' for an individual child to attend different schools based on the distance between their home and the schools. This represents the 'effort' in which parents may have gone to in order to overcome the 'friction' of the relative distances to alternative schools. Not only is the probability calculated on the basis of the distance between a child's home and a school, it is also based upon the relative distances to other alternative schools. Therefore, a pupil travelling five miles to their choice of school would have a greater probability of going to that school than another child also travelling five miles to the same school but who lived closer to other schools. These probabilities, then, provide a more accurate measurement of the likelihood that a child would attend a particular school in the context of the local education market. These measurements can then be used in numerous ways. For example, the average probability of attendance (between 0.0 and 1.0) for pupils attending each school reflects the popularity of these different schools; the lower the average probability score the more popular that school possibly is. Alternatively, the probability measures can be used to reflect the differing levels of 'activity' in the market place by parents and children; the higher the probability score for the school they do attend the less likely that they have been 'active' in the education market place. Furthermore, in combination with other socio-economic data (either collected for individual households or obtained from the area they live in) it is possible to examine the relationships between wealth and income, for example, with the levels of market activity as determined by the probability scores for individual households (Taylor 2002).

The last example illustrates another way postcode data can be used in a GIS to enhance our understanding of school choice and its impacts. In combining other datasets with the postcode data it is possible to ascribe key socio-economic (and other) variables to individual children, or households. Such data could be provided by the schools, such as their eligibility for free school meals or individual measures of attainment; obtained from individual children and parents, such as their attitudes towards education or the factors they deem important when choosing a school; or gathered from existing secondary data sets, such as the UK Census or other demographic datasets, for the local areas the families live in. Using point-in-polygon analysis the characteristics of the local area of individual households could be associated with the individual families in the study to enhance our knowledge about the households and the neighbourhoods they are located in. This is particularly helpful where we are examining a large number of families and schools, and where we are using the whole pupil population living in a local education market.

The third analytical capability of using GIS in the study of school choice is to examine in more detail the spatial–temporal dimensions to school choice and in particular the journeys to school. It has already been reported that the daily routines of travelling to school, the decisions about those journeys (Koppelman and Lyon 1981) and the way they are managed by parents (Valentine 1997), can affect children's use of space (Joshi et al. 1999). And it has already been shown that these travel decisions can constrain school choice in a variety of ways and for different people (Taylor 2002). However, using GIS to analyse journeys to school is the least developed of its technical capabilities. For example, it is possible to track and spatially reference the exact routes of individual children in travelling to school, using a GPS, to consider the spatial and temporal dimensions of those journeys. These could then be related to the decisions about school choice and to the use of, and attitudes towards, particular spaces by children. This would also contribute to our general understanding of mobility amongst young people (Pooley et al. 2005). An example of the technical capabilities of using GIS to study spatial–temporal mobility can be found in the work of Kwan (2000; Kwan et al. 2003). It is only a matter of time before such techniques are applied to the study of school choice.

CONCLUSIONS

This chapter has discussed and illustrated the use and benefits of GIS as a spatial research tool in the particular policy analysis of school choice and education marketization. Such an approach is also conducive to other studies of education policy. For example, the use of GIS can aid the examination of neighbourhood effects on school and individual attainment (Martin and Atkinson 2001; Openshaw and Turton 2001). Alternatively, it could be used to consider other examples of spatial patterns of participation in education, such as in higher education (Taylor and Gorard 2001b).

While the findings and examples presented here clearly do not cover all the issues relating to school choice it should now be clear how and why a geographical analysis of school choice provides a useful framework in which to continue to explore these issues. The developments in GIS and the availability of suitable data provide the means for further research in, and refinement of, spatial analyses of school choice. The particular advantages of using GIS in this area of education policy and in other areas would seem to be four-fold. First, it allows for a systematic analysis of the spatial dimensions to education policy; including the fragmented and localized delivery of national policy and the role that space and place can play in moderating and mediating the impacts of education policy. Second, using GIS as a spatial analytical tool accommodates complementarity with other data (through spatial analysis) and with other research (by producing generalizable findings from localized studies). Third, it adds a new analytical dimension to our theoretical and explanatory models, by providing additional comparators, i.e. across space, in a similar fashion to the ways in which analyzing trends over time can add greater analytical power. The fourth main advantage of using GIS and its associated spatial tools is that relatively large-scale analyses can be undertaken, something that can often be absent in many critical analyses of education policy. Not only does this mean that the design of smaller, more detailed, studies can be better informed, but it also provides the context or background in which other processes, decisions, and attitudes are located. This is important in ensuring that the potential structures and constraints to the actions and agency of individual families, parents, and children are appreciated.

REFERENCES

Adler, M., Petch, A. and Tweedie, J. (1989) *Parental choice and educational policy*, Edinburgh: Edinburgh University Press.

Ball, S. J., Bowe, R. and Gewirtz, S. (1995) 'Circuits of schooling: a sociological explanation of parental choice in social class contexts', *Sociological Review*, 43.1: 52–78.

Bourdieu, P. and Passeron, J.C. (1977) *Reproduction: in education, society and culture*, London: Sage.

Bradford, M. (1990) 'Education, attainment and the geography of choice', *Geography*, 75.1: 3–16.

Bradford, M. (1993) 'Population change and education: school rolls and rationalisation before and after the 1988 Education Reform Act', in T. Champion (ed.), *Population matters: the local dimension*, London: Paul Chapman.

Clarke, G. and Langley, R. (1996) 'A review of the potential of GIS and spatial modelling for planning in the new education market', *Environment and Planning* C, 14: 301–323.

Dore, C. and Flowerdew, R. (1981) 'Allocation procedures and the social composition of secondary schools', *Manchester Geographer*, New Series, 2.1: 47–55.

Giddens, A. (1984) *The constitution of society: outline of the theory of structuration*, Cambridge: Polity.

Glatter, R. (1989) *Educational institutions and their environments: managing the boundaries*, Milton Keynes: Open University Press

Gorard, S., Taylor, C. and Fitz, J. (2003) *Schools, markets and choice policies*, London: RoutledgeFalmer.

Heywood, I., Cornelius, S. and Carver, S. (1998) *An introduction to geographical information systems*, Harlow: Longman

Higgs, G., Webster, C. J. and White, S. D. (1997) 'The use of geographical information systems in assessing spatial and socio-economic impacts of parental choice', *Research Papers in Education*, 12.1: 27–48.

Jackson, B. and Marsden, D. (1966) *Education and the working class*, Harmondsworth: Penguin.

Joshi, M., MacLean, M. and Carter, W. (1999) 'Children's journey to school: spatial skills, knowledge and perceptions of the environment', *British Journal of Developmental Psychology*, 17: 125–139.

Kaufmann, V. (2002) *Re-thinking mobility*, Burlington: Ashgate.

Kaufmann, V., Bergman, M. M. and Joye, D. (2004) 'Motility: mobility as capital', *International Journal of Urban and Regional Research*, 28.4: 745–756.

Koppelman, F. and Lyon, P. (1981) 'Attitudinal analysis of work/school travel', *Transportation Science*, 15.3: 233–254.

Kwan, M. (2000) 'Gender differences in space-time constraints', *Area*, 32.2: 145–156.

Kwan, M., Janelle, D. G. and Goodchild, M. F. (2003) 'Accessibility in space and time: a theme in spatially integrated social science', *Journal of Geographical Systems*, 5.1–3.

Martin, D. (1996) *Geographical information systems: socio-economic applications*, London: Routledge.

Martin, D. and Atkinson, P. (2001) 'Investigating the spatial linkage of primary school performance and catchment characteristics', *Geographical and Environmental Modelling*, 5.1: 67–83.

McMillan, T. (2005) 'Urban form and a child's trip to school: the current literature and a framework for research', *Journal of Planning Literature*, 19.4: 441–456.

Openshaw, S. and Turton, I. (2001) 'Using a geographical explanations machine to explore spatial factors relating to primary school performance', *Geographical and Environmental Modelling*, 5.1: 85–101.

Pooley, C., Turnbull, J. and Adams, M. (2005) *A mobile century? Changes in everyday mobility in Britain in the twentieth century*, Aldershot: Ashgate.

Raper, J. F., Rhind, D. W. and Shepherd, J. W. (1992) *Postcodes: the new geography*, Harrow: Longman.

Reay, D. and Lucey, H. (2000) 'Children, school choice and social differences', *Educational Studies*, 26.1: 83–100.

Soja, E. (1980) 'The sociospatial dialectic', *Annals of the Association of American Geographers*, 70: 207–225.

Taylor, C. (2001a) 'The geography of choice and diversity in the 'new' secondary education market of England', *Area*, 33.4: 368–381.

——— (2001b) 'Hierarchies and "local" markets: the geography of the "lived" market place in secondary education provision', *Journal of Education Policy*, 16.3: 197–214.

——— (2002) *Geography of the 'new' education market*, Aldershot: Ashgate.

Taylor, C. and Gorard, S. (2001a) 'The role of residence in school segregation: placing the impact of parental choice in perspective', *Environment and Planning A*, 30.10: 1829–1852.

——— (2001b) 'Student participation of higher and further education in Wales', *The Welsh Journal of Education*, 10.2: 55–71.

Valentine, G. (1997) '"My son's a bit dizzy." "My wife's a bit soft": gender, children and cultures of parenting', *Gender, Place and Culture*, 4.1: 37–62.

White, P., Gorard, S., Fitz, J. and Taylor, C. (2001) 'Regional and local differences in admission arrangements for schools', *Oxford Review of Education*, 27.3: 317–337.

Wilson, P. R. and Elliott, D. J. (1987) 'An evaluation of the Postcode Address File as a sampling frame and its use with the OPCS', *Journal of Royal Statistical Society A*, 150: 230–240.

6 Disability, education, and space

Some critical reflections

Felicity Armstrong

There has been a recent blossoming of interest in cross-disciplinary approaches to social research. Ideas, concepts, discourses, and metaphors are clambering over disciplinary boundaries at such a rate that it is hard to know where some of them originally 'belonged'. This cross-fertilization of ideas is changing the face of some aspects of research in education. In particular, ideas from social and cultural geography are opening up fresh vantage points from which to examine familiar areas such as processes of selection and policy making in education. Soja observes that:

> spatiality and the inquisitive spatial imagination have recently entered, as a vital third mode of practical and theoretical understanding, that has hitherto been seen as an essentially two-sided socio-historical project. (Soja 1996: 6)

'Education' as a field of research has traditionally been approached as a project in which the 'spatial' is entirely absent, and historically the principal references to issues concerning space relate to technical-bureaucratic arrangements such as statutory guidelines regarding, for example, the physical arrangements of schools and the organization of school populations (see for example Porter 1906; DfEE 1996). Such arrangements—while presented as technical and rational, are, of course, deeply implicated in the production and reproduction of identities, difference, and power relations—but the explanatory force of the 'inquisitive spatial imagination' has been absent.

In this chapter I will explore some ways in which ideas about space and place are potentially powerful in transforming the way we can understand processes of exclusion and inclusion in education, particularly in relation to children and young people constructed as 'other' on the grounds of difference through the policies, discourses, and structures which make up what are referred to as 'education systems'. I will briefly consider examples of marginalizations which take place on the basis of culture and class as a means of exploring spatializations informed by 'geographies of power'. I have taken an example of the spatialization of class struggle as represented by the history of the Cutteslowe Walls in Oxford, and the 'spacing out' of Traveller/Roma Gypsies, as a way of making connections between familiar and readily recognized struggles over space (the physical divide between working class and middle class housing estates, the eviction of Traveller/Roma Gypsies), and some of the less familiar, or hidden, spatializations and struggles which relate to disability and education. I have drawn on some ideas from the work of Sibley, Lefebvre, and Soja, and in particular on Soja's notion of 'trialectics' and Thirdspace, as contributing to a possible theoretical framework for the discussion.

THE ENTRY OF THE SPATIAL

In earlier work I paid particular attention to 'historical' and wider political contexts as a means of exploring disability and difference and educational policymaking (e.g. Armstrong 2001, 2002). I became aware of the relationship between the exclusion of disabled children from ordinary educational settings and the impact of the built environment, the organization of space and the values which underpin demarcations, re-partitions, and the management of educational structures and curricula. I began to make connections between the imposition of a centralized curriculum in the UK with the ironing out of cultural differences (Armstrong 1998) and other kinds of exclusions. I drew parallels between the generation and regeneration of social differences through procedures and processes relating to curricular selections and the divisions and organization of students for learning, and the way schools are designed with their spaces divided up and allocated to differentiated groups and activities, allowing 'social and cultural reproduction to be organized' (Armstrong 1999: 75). I read Imrie's work and began to understand more about 'the spatial', and the way in which 'the built environment is a social and cultural product' (Imrie 1996). I also read Sibley's book *Geographies of Exclusion* and became interested in the 'processes of boundary erection', the creation of separate and separating places, and the production of stereotypes. (Sibley 1995). It seemed to me that many dimensions, whole galaxies of concepts and experience, had been missing from my thinking in the past, which I have now begun to explore (Armstrong 2003).

My notebooks from the late 1990s record some 'critical moments' in terms of the development of ideas and experience. One example will illustrate such 'moments'. I read Collison's book *The Cutteslowe Walls*, published in 1963, which tells the extraordinary story of the building of two nine-foot high walls by a private developer, in Oxford in 1934, across the roads which linked his private estate to an adjoining council estate. In spite of attempts to physically pull down the walls on the part of residents of the council estate and their supporters, the walls remained in place for 25 years, during which time the residents of the council houses had their route blocked to shops, buses, and schools and were forced to take a long detour if they wished to leave their imposed enclave.

Johnston (1991) insists on the importance of recognizing the *particularities* of place by appreciating the nature of society attached to particular sites. The Cutteslowe Walls were the embodiment of class struggles and social relations in Oxford at the time. This was a period of rapid growth in the car industry, with an ever-increasing demand for cars on the part of the wealthy citizens of Oxford. There was a need for cheap housing to accommodate the growing numbers of motor industry workers and their families—but social housing and working class communities were, by and large, to be found to the east of the city within easy reach of the factories, and well away from the leafy university suburbs of north Oxford. Those who lived on the Cutteslowe estate were 'out of place' and, in a profound sense, 'trespassing'. The walls were both a physical barrier to would-be transgressors, a symbol of power relations and social struggles and a message about territory. Thus, the walls did not only *stand for* class difference and struggle, snobbery, an arrogant colonizing territorialism, and a *mépris* for local working class people—expressed in the disregard for their right to move freely between their homes and places of work, schools, and shops—but played an active part in the construction and affirmation of class differences, power relations, and differentiated territories. Attached to these 'territories' were revised or magnified identities of subjugated 'worker-tenant' and powerful 'private home owner'. Places, which had common thoroughfares, became no-go areas. Children from both sides of the walls no longer crossed paths with their 'neighbours' but were coerced along different routes on their way to different schools. As Lefebvre explains:

> The relationships established by *boundaries* are certainly of the greatest importance...along with the relationship between boundaries and named places.... Every social space, then, once duly demarcated and oriented, implies a superimposition of certain relations upon networks of named places, of *lieux-dits*...resulting in different kinds of space... 'accessible space for normal use,' 'boundaries and forbidden territories,' 'places of abode' and 'junction points' which are often 'places of passage and encounter'. (Lefebvre 1991: 193, original emphasis)

In the story of the Cutteslowe Walls the physical boundaries are easily identifiable, as are some of the consequences. It is useful in drawing attention to the way in which 'space is increasingly created as...commodified enclaves... within which at least parts of everyday life can be carried out in isolation from other social groups' (Thrift 1996: 131). The story also suggests that in the process of boundary erection, power struggles over other issues are taking place—in this case, over class, difference, the fear of proximity, and the right of free circulation. I have noticed that these are struggles which constantly re-emerge in different contexts and forms.

SPACE AND THE 'TRIPLE DIALECTIC'

Soja argues for the adoption of a 'triple dialectic' through which to explore the 'limitless dimensions of our social spatiality' by linking 'historicality, sociality and spatiality'. He uses Lefebvre's term 'transdisciplinary' to mean '...not being the privileged turf of such specialised fields as History, Sociology, and Geography, but spanning all interpretive perspectives' (Soja 1996: 6).

In general, there has been little recognition in educational research of transdisciplinarity and, in particular, the contribution of geography and architecture in terms of understanding education in social contexts. However, the work of Ball (Ball 2003; Ball et al. 1995) and Vincent (Vincent and Martin 2002), which provides a spatial reading of school choice and 'risk and class', suggests that space and place are crucial parts of the education-class nexus. Imrie has written extensively on the spatialized social relations and perceptions which contribute to, and emerge from, the marginalization of different sections of the population, and of disabled people in particular:

> social inequalities in the city, as one of the more significant spatial materialities, are mapped by the complex spatial mosaic of place-based segregation, like suburbanization, ghettoization and gentrification, and, in this sense, such forms of segregation are expressive of the culturally imperialist values of society. (Imrie 1996: 12)

The insistence on *values* and the way these make their way through the structures, spaces, and discourses of social life is crucial in understanding the implications of the way we organize the social world. As Cresswell (1996: 17) argues, 'the meanings of place are created through practice', and social practice is not value-free.

SPACE AND HEGEMONY

While the city is the arena in which 'place-based segregations' are most evident and concentrated, there are many different spatially articulated contests which are not specifically located in the city, but move around. Travellers, and Roma/Gypsies, are in a continuous state of near-siege across parts of Europe in terms of where they can make their homes, congregate, and pursue their culture and lifestyles, unmolested. One kind of official response has been the development of policy initiatives designed to incorporate and assimilate Travellers and Roma/Gypsies, especially through the establishment of controlled settlements in particular areas and the setting up of education structures and services. However, there have been other policy initiatives from different departments which seek to move them on and prevent them from becoming established near residential areas. A report from the Office for Standards in Education (OFSTED 2003) highlights a major contradiction in the ways policies are implemented at a local level. While the Traveller Education Service is engaged in finding places in schools that will accept Travellers, the Traveller liaison officer will be seeking to evict them from 'illegal' sites. Such struggles, which are spatially mapped, represent contrasting hegemonies within the same set of 'culturally imperialist values' of British society, namely:

1. As a project of normalization, education is a good thing, and a right, for all children and, in the case of Traveller and Roma/Gypsy children it will assist in integrating them into the wider society and 'lifting them out of poverty'.
2. Traveller and Roma/Gypsies must be controlled and prevented from 'trespassing'. They must not be allowed to live within or near to the community.

These twin, seemingly opposed, projects co-exist and provide a perennial focus of debate and argument in the media and in parliament and local councils. Yet they harmonize in their underlying shared belief, which is never challenged, that Travellers and Roma/Gypsies are potentially out-of-control outsiders who need to be managed either through assimilation and acculturation (requiring forced immobility) or forced exclusion and repudiation (requiring forced mobility).

Sibley (1995) has shown how groups are marginalized in society through particular constructions of space and 'imagined geographies' (Sibley 1995). The stereotype of the 'English countryside' is productive of 'discrepant others' who do not fit in, or worse, defile, the idyll of the English rural 'way of life'. Thus groups such as 'Gypsies, and Irish and Scottish Travellers who are not already accommodated on official sites, ravers, hunt saboteurs and environmental protesters' (Sibley 1995: 106–107) have become criminalized under the Criminal Justice and Public Order Act of 1994.

Similarly, the construction of the notion of the ideal school has been a cornerstone of New Labour education policy, creating hegemonic relationships and practices around 'standards' and 'performativity'. Schools must strive for excellence and pupils are to 'reach their potential' in an increasingly selective and hierarchically organized system, governed by evaluation procedures and controls in which 'failing' schools are shut down, and disturbing students removed to segregated sites such as 'Pupil Referral Units' or special schools. Such closures and removals are only possible because of the existence of what Sibley describes as 'fictional characterisations' which deny the heterogeneity of people and communities, cultures and learning styles, kinds of knowledge, physical shape and appearance, performance, and behaviours. Just as the fiction of the English rural 'way of life' constructs those who do not fit into this mythical landscape as deviant and disturbing, the 'fictional characteristics' of the 'ideal pupil' underpin education policy-making in ways which have excluded disabled children and those who did not 'fit in' to the idealized landscape of the mainstream school.

EDUCATION POLICY MAKING

There are dominant accounts of policy—routine in the mouthpieces of government—as rational, made through legislation and passed along through the structures and conduits of government, from central, to local, to institution. These accounts do not concern themselves with the ways in which policies are mediated and transformed by local social and economic conditions, resulting in the making of further, compensatory, 'corrective' or coercive policies to manage these. In some respects, many of the branches of New Labour education policy in the UK have been made to deal with possible 'fall-out' from their core policies which have done little to address extremes of wealth and poverty, and maintain and promote selections in education based on an acceptance of systems increasingly driven by the principles of the free market (Ball 2003; Gewirtz 2002).

The term 'systems' suggests boundedness, rationality, and homogeneity. Another way of thinking about 'systems' is that they are unbounded and irrational, amorphous, in a condition of permanent flux and reconfiguration, thrown up by social, geographical, cultural, and economic movement and spaces in which human beings, as individuals and groups, work and are worked upon. Some commentators have emphasized the multiple levels, arenas and social actors involved in the production of policies, and the relationship between policy processes and outcomes and discursive practices in forming and mediating policy (Hindess 1986; Fulcher 1989). It is necessary to move away from thinking about education and social relationships in terms of policymaking as a coherent system, and take account of the fluidity and spatiality of relationships. In exploring and drawing on theories from social and cultural geography in particular, it becomes possible to reinterpret assumptions about the ways in which relationships, places, and identities are formed, assigned, and reassigned, through processes of spatial ordering and reordering.

Through a critical examination of assumptions that different groups 'belong' in different places in some natural or historical sense, we can begin to see the social, political, and environmental factors which are involved through a spatial reading of policy as it emerges in ever changing formations. We cannot separate the spatial from the social (Massey 1994) because social roles and identities, relationships and practices, are produced by, and productive of the cultures, geographies, and social lives of places and communities. The dynamic relationship between the social and the spatial is at the heart of the emergence of places, communities, and identities. Thus, the 'special school' is literally a 'solid structure' in a material sense 'but also "solid" in the sense of cultural, pedagogical, political, economic structures' (Derrida 1989: 8). It both absorbs cultural meanings, and declares them. The special school is the confirmation of abnormality, an aspiration for mutual protection for those on the inside and those on the outside; it is a place for 'special' people, with exotic behaviours, appearances, and requirements. The space of the special school is both produced by, and productive of cultural meaning. It is a place hewn out of human practices peculiar to itself (discourses, therapies, curricula, kinds of professionalism, relationships of power and dependency) and from the wider environment which gives it shape and wider significance in relation to other kinds of places. The special school is inscribed with the symbolic meanings and social practices of, for example, 'special' education, health policies, and medicalized models of difference, market forces, and 'parental choice', dominant attitudes and beliefs, the organization of the physical environment and the messages of charitable organizations, and a generalized fear and distrust, mixed with curiosity, of the non-conforming other.

SPACE, PLACE, AND BINARY DISCOURSES

A spatial reading of the movement between education structures, policies, and practices can contribute to the development of a reflexive, critical stance towards them in which what Grace (1977) referred to as 'radical doubt' emerges as part of a revised understanding of the 'familiar and taken-for granted characteristics' of education. We need this radical doubt in order to think about education policy differently, especially in relation to processes of inclusion and exclusion in education. This may be seen as a move towards a cultural-political reading of policies as they interrelate with wider contexts and movements. Soja (1996) draws an important distinction between different established forms of a counter hegemonic cultural politics of difference:

> The first and most 'acceptable' is rooted in the post-Enlightenment development of liberal humanism and modernism. It has traditionally based its opposition on the assertion of universal principles of equality, human rights, and democracy, seeking to reduce to a minimum the negative effects of difference, whatever their origins. A second form... not always completely separable from the first, arises from more radical contestation over the many axes along which socially constructed power differentials have historically and geographically polarized. Rather than aiming above all to erase differences or to 'even things out,' it uses difference as a basis for community, identity and struggle against the existing power relations at their source. (Soja 1996: 88–89)

Among the axes we need to examine are those of space, place, and binary discourses. Soja observes that counter-hegemonic struggles have traditionally taken place around such categories of cultural consciousness:

> as class, race, ethnicity, nationality, colonial status, sexuality and gender which usually preserve inter-categorical distinctions while seeking to eliminate differences in power along each 'internal' axis. (1996: 89).

This has certainly been the case around the category of 'disability' and radical disability politics and activism. The spatial readings of historical and political struggle suggested by Lefebvre and Soja provide a radical alternative to the boundedness of struggles based around categories of difference. Interestingly, 'disability' is not included in the binary pairs of oppositions listed by Soja (1996): 'capital/labour, self/other, subject/object, colonizer/colonized, white/black, man/woman, majority/minority, heterosexual/homosexual'. But the binary oppositions of 'disabled/non-disabled', 'normal/not normal' are as much in need of critical deconstruction as others in order to reveal the 'social and spatial construction of difference as a means of producing and reproducing systematic patterns of domination, exploitations and subjection' (p. 90). As Imrie (1996) argues:

> dualistic thinking tends to hide the complex and contested socio-cultural processes and practices which underpin the construction of dualities and...there is no easily defined state of being 'disabled' or 'ablebodied' in that such categories are constantly being challenged, contested, and transformed in wider socio-political and institutional practices.... Part of such processes relates to the interaction between 'states of disability' and geographical specificity. (1996: 13)

The space of the special school, and the cultural distinctiveness and location of different types of schools, which are reproductive of categories of difference, power relations, and values in the wider society, require a spatial reading in order to understand the role that human geographies and spatializations play in constructing and confirming dualities, and hardening categories. Embedded in these processes and the spaces they create are the distinctive discourses belonging to particular arenas.

LEFEBVRE AND SOJA

The space of the school may be described in similar terms to those used by Lefebvre to describe the city as 'fashioned, shaped and invested by social activities' (Lefebvre 1991). In *The Production of Space* (Lefebvre 1991) Lefebvre developed his triad, or 'three moments' of social space: 'spatial practice', 'representations of space', and 'spaces of representation'. Briefly, 'spatial practice'—or 'perceived space'—concerns space as it is interpreted and referred to in commonsense ways (e.g. 'home', 'school') and the spatialized social routines of everyday life. 'Representations of space' or 'conceived space' refers to the more abstract notion of space as used in maps, town planning, and the 'means of production' and the ordering of the environment signalled by 'signs and codes'. In this 'storehouse of epistemological power' (Soja 1996: 67) discourse and texts serve as regulatory and declamatory markers. The third in the triad—'representational spaces' or 'spaces of representation'—refers to space as 'directly lived through its associated images and symbols, and hence the space of "inhabitants" and "users"' (Lefebvre 1991: 39).

Soja built on Lefebvre's work to develop a 'trialectics of spatiality' which distinguishes between the concept of 'Firstspace' as something observable and measurable, and 'Secondspace' which, Soja explains, is:

> entirely ideational, made up of projections into the empirical world from conceived or imagined geographies. This does not mean that there is no material reality, no Firstspace, but rather that the knowledge of this material reality is comprehended through thought. (Soja 1996: 79)

Soja's 'Thirdspace' suggests a space of critical deconstruction of the duality of the 'Firstspace' and 'Secondspace', inviting a challenge to the privileging of either a framing of space as entirely material, or as purely subjectively constituted. Thirdspace proposes a space in which:

> *Everything* comes together...subjectivity and objectivity, the abstract and the concrete, the real and the imagined, the knowable and the unimaginable, the repetitive and the differential, structure and agency, mind and body, consciousness and the unconscious, the disciplined and the transdisciplinary, everyday life and unending history. Anything which fragments Thirdspace into separate specialised knowledges or exclusive domains—even on the pretext of handling its infinite complexity—destroys its meaning and openness. (Soja 1996: 56–57, original emphasis)

The last part of this quotation seems to warn off attempts at deciphering the 'constituents' of Thirdspace. However, the notion of Thirdspace may provide an exciting lens through which to critically explore the spatialities of education and disability. The ideas outlined above are useful for seeing, from fresh perspectives, aspects of what is involved in the spacing and placing of children within educational structures. They can be used to challenge, in particular, the idea that exclusions of disabled children from ordinary schools—still the dominant practice in most countries, including many local authorities in the UK—can be explained either in terms of the barriers presented by the built environment (e.g. purely material) or in terms of subjective conditions, including attitudes, expectations, imagined differences, and constraints. Thirdspace would embrace both these explanatory strands, but would envisage a far more complex, dynamic, space in which there is constant change and flux and fragmentation informed by material conditions and constraints, local cultures, individual beliefs, power struggles, competing interests and discourses, movement, and shifting understandings. I am reminded of Ball's description of 'policy' as:

> both text and action, words and deeds, it is what is enacted as well as what is intended.... Policies are always incomplete insofar as they relate to or map on to the 'wild profusion' of local practice. (Ball 1994: 10–11)

And as:

> representations which are encoded in different ways (via actors' interpretations and meanings in relation to their history, experiences, skills, resources and context). A policy is both contested and changing, always in a state of 'becoming,' of 'was' and 'never was' and 'not quite'. (Ball 1994: 16)

Ball's characterization of policy resonates with the dynamic multiplicity, discontinuity and constant movement suggested by the notion of Thirdspace in which 'everything comes together'.

In the following section I shall try out a re-examination of a research context taken from an earlier qualitative study (undertaken with Tony Booth in the early 1990s) through the lens of Soja's 'trialectics'. The choice of this research, rather than another one, has been made because of the very distinctive spatializing behaviours and practices it reveals, although I believe that, in theory, a 'trialectic' reading of policies and practices in education should be possible in any situation.

THE STORY OF THE 'SECURE DOORS'

A 'critical moment' during the research process of a particular evaluation project made me acutely aware of the fundamentally 'spatialized' nature of educational arrangements and, in particular, as they relate to disabled people. In this case, I was doing fieldwork in a large comprehensive school in the north of England. The school had recently expanded as a result of the closure of a special school in the region, which had been designated for students 'with severe learning difficulties'. A whole new 'suite' had been built to 'accommodate' the new students from the special school, with three classrooms, different rooms for various therapies, a kitchen and dining area, a hydro-pool, and a 'social' area with provision for a cafeteria run by the students. Selected students from the main school came to the café on a rota basis to 'help' and to 'socialize'.

For the first time, as part of a stated commitment, on the part of the local education authority, to developing inclusive education, disabled students set off in the morning to the same school as other young people in the area, rather than travelling in the opposite direction to a beaten-up Victorian building which had been adapted in an ad hoc way to accommodate disabled people. The comprehensive school was also accessible through the main entrance, but a separate accessible entrance had been added to the newly built suite, behind the complex of buildings, in order to 'reduce the risk of congestion which might arise if wheel chairs used the main entrance'. Other separating practices were put in place, based on perceptions of difference. I found that students who required assistance with eating ('feeding' was the word used by a professional) were to have their lunch separately in a specially designed eating area. Other disabled students would eat in the main dining area, but those using wheel chairs would eat 'slightly before' other students so as to avoid 'blocking passage ways' at busy times. Perhaps the most startling discovery was that the new suite was separated off from the rest of the school by three sets of 'secure' doors which could not be opened by anybody in a seated position (i.e. from a wheel chair) or, indeed, anyone not able to use both hands. When I asked a senior teacher about the doors, it was explained that they were designed to stop people just 'wandering around' and, in particular, to ensure that people didn't 'just drift in' from the main school. Another teacher, who had worked in the special school, said:

> The double doors prevent many of our students from circulating freely.... The sixth-form block is completely isolated...and students have actually got to go through three double doors to the rest of the block. We have three students in wheelchairs who would be able to get themselves there independently if there were no doors. In the old school, where all the classrooms were arranged around the hall, the students were more independent. Here their independence has gone. They've got to be pushed everywhere and sometimes you need two staff so one can open the doors. The students are aware of that. Debbie, for example, has said, 'I like the new school but I can't get around,' so she's aware that it has put constraints on her. (Armstrong and Booth 1993)

What contribution does a spatial reading of the scenario described above and, in particular, Soja's 'trialectics', make to improve our understanding of what is happening? First, there are the physical, organizational characteristics of the main school and the new suite; the physical environments of the two sites have become familiar and hence 'known' and can provide some kind of shared (albeit uneven) reference point in the school community. This, and the physical connections and barriers between the two areas, might be mapped onto Soja's 'Firstspace' in which the 'space' is knowable and measurable in a tangible way.

In 'Secondspace', the focus is on conceived rather than perceived space in which 'spatial knowledge is primarily produced through discursively devised representations of space, through the spatial workings of the mind' (Soja 1996: 79). If we apply this notion to the example above, as encompassing purely 'ideational...projections into the empirical world from conceived or imagined geographies' we can look at some usually taken-for-granted social roles and activities from a different angle.

Consider the work of architects, planners, advisors, and professionals with their individual 'imagined' pictures of what the space of the reconfigured school would 'look like'—a space which was to join 'others' onto an existing school, a new project, a new group of people—disabled children and their teachers, support staff, and therapists, specialist equipment and installations. What might have been the images, words, and imagined geographies they drew on in thinking and talking about the project? What ideas about disability, difference, need, security, community, and the meaning of 'school' would inform their imaginations and deliberations, and what kind of architectural plans, financial forecasting, and arrangements would be drawn up and presented to committees as a result? At this point, the relationship between First and Secondspace becomes clearer, but much is left out in terms of the complexities involved, because we also need to consider what notions of disability and difference, what values and experiences, informed the individual and collective 'projections into the empirical world from conceived or imagined geographies' of those charged with planning. This opens up the creative possibilities provided by 'Thirdspace'—or spaces of representation 'in which the real, the imaginary and the hyper-real coexist without one being prioritized over the others' (Soja 1997: 249). 'Thirdspace' provides an opportunity for exploring processes of inclusion and exclusion unencumbered by bounded mono-concepts such as 'physical barriers', 'categories of impairment', 'attitudes', 'mythologies', 'discourses', 'professionals', 'power', 'legislation', and so on. These are all *there* but they are part of, and in constant interaction with, the multiple dimensions of the 'real' and the 'imaginary'.

A closer look at the micro-politics of the daily life of the school and its adjoining newcomers revealed some of this multiplicity. The planning and organization of the school as described had many dimensions and consequences. One was the increasing reluctance on the part of teaching staff to establish practices which would enable students to participate in lessons with their peers, and share recreational time. An additional problem was the fact that it was understood that disabled students would always have one of 'their' teachers with them if they attended mainstream lessons. This would frequently reduce the staffing levels in the 'suite' to an unacceptable level for other, remaining students. Another issue was concern expressed that if students became involved in complex timetable arrangements with the main school, they would miss important therapy sessions and other activities. Furthermore, disabled students did not share mealtimes or recreation periods with their peers in the main school: it was considered too difficult to organize, and students were, in general, unable to circulate freely and thereby choose between the different recreational activities and areas. The assumption that disabled students would necessitate additional, specialist staff in the recreation areas—although it was never made clear what the basis was for this argument—was a further reason why disabled students invariably spent their recreation time inside, in the 'cosy atmosphere' of the specialist 'suite'. The further assumption that disabled young people, and their wheelchairs, would constitute a safety hazard, was never examined and consequently the possibility of a basic social activity such as the sharing of a meal at midday, was ruled out. Here we have a collection of scenarios which are deeply veined with dominant assumptions, fears, ideas about professional roles, constructed images of the 'other', the privileging of bureaucratic and procedural concerns over a commitment to listening to the voices of all concerned, and confusion over the very nature about the meaning of 'community' and 'school'—all coming together in a Thirdspace.

The readily recognizable spatializations in this scenario, informed a space in which '*everything*' came together, but not always in easily visible and tangible forms, such as: the subjective interpretation and expectations of the duality—disability and 'the normal', the conscious and unconscious fear of difference based on the 'knowable' and the 'unimaginable', the cumulative effects of 'daily life' and all the *embedded knowledge* about who belongs where, and mythologies of difference with their 'unending histories'. In the situation above, the segregated space of the special school had been recreated within new boundaries—those of the community school. But the social space of the newly added suite built to 'accommodate' (contain) disabled students, and their therapists, teachers, and support staff, was also bounded by physical barriers, practices, and deeply embedded discourses and assumptions about being and belonging, not being and not belonging. The signs and signals displayed by the three sets of double doors, the separate entrances and recreational and learning spaces, the paraphernalia of the therapeutic, and the material manifestations of fears about the possible disorder and danger created by allowing people to be 'out of place', all contributed in tangible and intangible ways to a powerful project of segregation, rather than inclusion.

CONCLUSIONS

In this chapter I have begun to explore some 'spatial' approaches to examining the structures and processes through which people, and children in particular, are 'spaced out' of 'mainstream' settings. I have tried to challenge the notion that linear, uni-disciplinary approaches are sufficient to explain processes of marginalization and segregation. The contribution of ideas from social geography and, in particular, a geography which itself is open and seeking out perspectives from other disciplines, highlights what a great deal of work we have to do in terms of exploring and decoding the deep movements and multiple dimensions and spaces of exclusionary forces.

One purpose of this chapter was to experiment with ideas about processes of spatialization which are still unfamiliar in the realm of research and debates in education. I wanted to see what happens when we make disciplinary boundaries more porous and challenge traditionally embedded binaries relating to normality and difference from unusual angles. In the second part of the chapter I returned to an earlier piece of research and tried to understand it from a fresh vantage point— through a spatial reading of events and practices. To do this, I borrowed from Soja's 'trialectics' to provide the beginning of a useable outline for a theoretical framework. In particular, the idea of Thirdspace opened up creative opportunities for challenging familiar and linear monochrome expositions and critiques of education and disability which focus on, for example, policy and the history of legislation, notions of disability, arguments about resources and cost-effectiveness, or charitable models of need and dependency, the 'rights' of disabled children and the 'rights' of 'mainstream' children. Thirdspace is an open terrain in which *everything comes together*—the historical, the political, the personal and private, the public and the legislative, the competing demands of the human embrace in the face of human fear, the collective and the individual, the spatial and the territorial, professional surveillance, and the persistence of the struggle for freedom and justice (in so many different spaces and configurations), the interplay of myth, memory, and imagination. How *could* one discipline such as 'psychology,' or one 'field' such as 'education' or 'educational management' ever begin to contemplate or explore all the possible levels, relationships, and biographies which touch, for example, on 'The Story of the Secure Doors'?

The objective of this chapter, therefore, has been to experiment with using some possibly unfamiliar theoretical ideas as a stimulus for thinking creatively about movements between policies, processes, and instances of inclusion and exclusion in a fresh light, from different vantage points, to provide a counter-perspective of social life as a process of change. When we use theoretical ideas from another discipline, these can *only* be (i.e. can never be more than) stimuli for thinking creatively. Theory has to help to understand, or explore. Disciplinary borders become infused with new ideas, or ways of seeing, creating fresh ways of thinking. Of course, the aim of using theoretical ideas from 'another discipline' is not to routinely apply them to a familiar discipline. They may not work. The aim is rather to let these ideas spark a new understanding of the 'familiar'.

ACKNOWLEDGEMENTS

I would like to thank Kalervo Gulson and Colin Symes for their critical commentary and helpful suggestions which they made during the process of writing this chapter.

REFERENCES

Armstrong, F. (1998) 'The curriculum as alchemy: school and the struggle for cultural space', *Curriculum Studies*, 6.2: 145–160.

—— (1999) 'Inclusion, curriculum and the struggle for space in school', *International Journal of Inclusive Education*, 3.1: 75–87.

—— (2001) 'Intégration ou inclusion? L'évolution de l'éducation spéciale en Angleterre', in *La Revue Française de pédagogie*, 134.

—— (2002) 'The historical development of special education: humanitarian rationality or 'wild profusion of entangled events?' *History of Education*, 31.5: 437–456.

—— (2003) *Spaced out: policy, difference and the challenge of inclusive education*, London: Kluwer Academic.

Armstrong, F. and Booth, T. (1993) *Learning for all: unit 16*, Milton Keynes: Open University Press.

Ball, S. J. (1994) *Education reform: a critical and post-structural approach*, Buckingham: Open University Press.

—— (2003) 'The risks of social reproduction: the middle class and education markets', *The London Review of Education*, 1.3: 163–175

Ball, S. J., Bowe, R. and Gewirtz, S. (1995) 'Circuits of schooling: a sociological exploration of parental choice of school in social class contexts', *The Sociological Review*, 43.1: 52–78.

Cresswell, T. (1996) *In place/out of place: geography, ideology and transgression*, Minneapolis: University of Minnesota Press.

Department for Education and Employment (DfEE) (1996) *The 1996 School Premises Regulations*, Circular 10/96. London: DfEE.

Derrida, J. (1989) 'Jacques Derrida in discussion with Christopher Norris', in A. Papadakis (ed.), *Deconstruction 11*, London: Academy Editions.

Fulcher, G. (1989) *Disabling policies? A comparative approach to education policy and disability*, London: Falmer Press

Gewirtz, S. (2002) *The managerial school: post-welfarism and social justice in education*, London: Routledge.

Grace, G. (1977) 'The "problem" of the urban school: some radical and Marxist formulations', in J. Raynor and E. Harris (eds.), *Schooling in the city*, London: Ward Lock Educational.

Hindess, B. (1986) 'Actors and social relations', in M. L. Wardell and S. P. Turner (eds.), *Sociological theory in transition*, Boston: Allen and Unwin.

Imrie, R. (1996) *Disability and the city: international perspectives*, London: Paul Chapman.

Johnston, R. J. (1991) *A question of place*, Oxford: Blackwell.

Lefebvre, H. (1991) *The production of space*, Oxford: Blackwell.

Massey, D. (1994) *Space, place and gender*, Cambridge: Polity Press.

OFSTED (2003) *Provision and support for traveller pupils* (Document reference number: HMI 455) London: Office for Standards in Education.

Porter, C. (1906) *School hygiene and the laws of health*, London: Longmans, Green and Co.

Sibley, D. (1995) *Geographies of exclusion: society and difference in the West*, London: Routledge.

Soja, E. W. (1996) *Thirdspace: journeys to Los Angeles and other real-and-imagined places*, Oxford: Blackwell.

—— (1997) 'Planning in/for postmodernity', in G. B. Benko and U. Strohmayer (eds.), *Space and social theory: interpreting modernity and postmodernity*, Oxford: Blackwell.

Thrift, N. (1996) *Spatial formations*, London: Sage.

Vincent, C. and Martin, J. (2003) 'Class, culture and agency', *Discourse*, 23.1: 109–28.

7 Working the in/visible geographies of school exclusion

Pat Thomson

The UK government has made much of its programme of bringing order to schools and neighbourhoods. It has introduced a veritable pantechnicon of policies: these range from the much publicized anti-social behaviour orders (ASBOs) issued to young people at loose in public housing estates and the prosecution of parents whose children fail to attend school, to a renewed focus on 'joining up' health, welfare, and education services to better integrate provision. A recent ruling also provided safeguards for children and young people under the age of 16 years who were permanently excluded from school—they had to be back in some form of education and training within 15 days from the date of their exclusion.

This requirement might have closed the gap on any head teachers who played fast and loose with local authority attempts to compel them to enrol students from other schools. But legislating for a speedy return to school does not deal with disincentives to do so. There are considerable pressures on all schools to attract enrolments by virtue of their well-behaved student population. There are also punitive consequences for schools which do not continually improve overall performance on tests and inspections, something difficult to achieve with significant numbers of students who are on the edges of attendance and acceptable behaviour. UK schools are caught in a pincer movement between standards and inclusion.

This situation is not unique to the UK. The tension between performance, popularity, and official approval and catering for a diverse group of students, not all of whom add to reputation or results, also exists in other school systems. In the United States, it is often tangled in the politics of de-tracking and is strongly racially inflected (Lipman 1998; Oakes 2005; Rubin 2003). In Australia, it is woven tightly into the fierce competition between public and private schools for funds and status (Reid 1998; Campbell and Sherington 2006).

In this chapter I examine the politics and practices of school exclusion and then move from critique to point out to some options that may produce different outcomes for students. While there is a considerable literature on school exclusion and inclusion (e.g. Brodie 2001; Cullingford 1999; Munn and Lloyd 2005; Osler and Vincent 2003; Parsons 1999; Pomeroy 2000;

Vulliamy 2001; Wright, Weekes and McGlaughlin 2000), very little of it foregrounds spatial practices (see also Armstrong 2003; Moss and Petrie 2002). I focus here particularly on three alternatives to exclusion: (1) resistance (de Certeau 1988) and conformity (Bhaba 1984; Szkudlarek 1993); (2) changes in place; and (3) the construction of counter-public spaces (Fraser 1997). My concern is how to make policy possible within the constraints outlined above.

In making the case for troubling the in/visible geographies of exclusion, I read across a number of research projects and locations, but use as the prime source and location a recent evaluation of a multi-site project intended to avert permanent exclusion and give students a second chance at a new school (Harris, Vincent, Thomson and Toalster in press; Thomson, Harris, Vincent and Toalster 2005). I also mobilize data from an Australian action research project which demonstrably shifted the educational fortunes of a group of young women apparently destined for early school leaving (Thomson, McQuade and Rochford 2005a, 2005b).

I begin with some comments on the spatial orientation taken in this chapter (in particular drawing on Lefebvre 1991; Massey 2005; Sibley 1995; Soja 1999) and then go on to sketch how excluded students experience the modern English school.

AN ORIENTATION TO THE SPATIAL

This chapter is situated within the emerging body of educational research which draws on cultural geographies (e.g. Fine and Weis 1998; Gordon, Holland and Lahelma 2000; Leander and Sheehy 2004; Nespor, 1997; Paechter, Edwards, Harrison and Twining 2001). Rather than seeing schooling as occurring within a fixed container and/or as a static, mappable geography, this body of scholarship is underpinned by notions of space as productive and dynamic. In particular this corpus holds that:

- Space is constituted through social relations and is also constitutive of them (Lefebvre 1991).
- Spatial relations are materially embedded and enmeshed in cultural-economic-political practices. These are not simply local but extend beyond the school to take in local/global flows, trajectories, and networks (Allen, Massey and Cochrane 2000). School cannot be seen as an impervious 'black box'.
- Spatiality cannot be separated from temporality. Spatial relations develop not only over time, but in time; identities and entities (institutions and nations, for example) are not formed prior to, or after the establishment of spatial relations but are co-constructed with, in, and through them (Massey 1994). Spatial relations must be conceptualized

as always in formation and never entirely closed. Educational geographies can therefore be thought of as an ongoing accomplishment.
- Space is multi-dimensional. According to Lefebvre (1991), there is a trialectic of (a) perceived space—the patterned routines of material, social practice; (b) conceived space—ideological, symbolic and representational practices; and (c) lived space—the everyday. When thinking about schooling, it is important to consider not only the material and the symbolic but also the realm of experience. Contradictions in and between perceived and conceived space are experienced passively but they are *actually* changed in lived space.

Space does not over-determine the social and the material. As Soja (2004: x) puts it, 'If our spaces and places, our human geographies, are socially constructed, it logically follows that they are not immutable or naturally given. This means that they can be socially changed, made into something better than they were through collective action'. Educational research that works for social justice (see Griffths 1998) thus should include a focus on the possibilities for change in and through troubling and shifting spatial relations.

Space is not homogeneous. Recognition that there are multiple experiences, trajectories, and narratives of space and place produces more open/less closed theory and practice. Thinking and acting spatially in this way creates the opportunity for a 'sphere of possibility' in which there is 'coexisting heterogeneity', multiplicity, and difference (Massey 2005: 9).

This chapter aims to demonstrate what this theorization of and focus on spatialization might mean for changing the everyday experiences of young people who face exclusion from school. In accord with Massey's urgings to focus on difference, I begin my exploration of exclusion with representations of the perceived and lived spaces of students on the margins of education.

THE GEOGRAPHIES OF EXCLUSION

Exclusion and perceived space

Perceived space is the readily visible and apparent. It is the material manifestation of social practice. Students must operate within a modernist educational organization characterized as 'spaces of enclosure' (Lankshear, Peters and Knobel 1996). This can be best represented as a time–space grid (see Figure 7. 1).

The grid could equally be a school diary, a timetable, graph paper on which a bell curve of test results might be plotted, or the national curriculum. As timetable, the boxes suggest lessons—the teacher and students can move within the confines of the lesson time—but not transgress the borders, go over or under time, because that will disrupt all the other

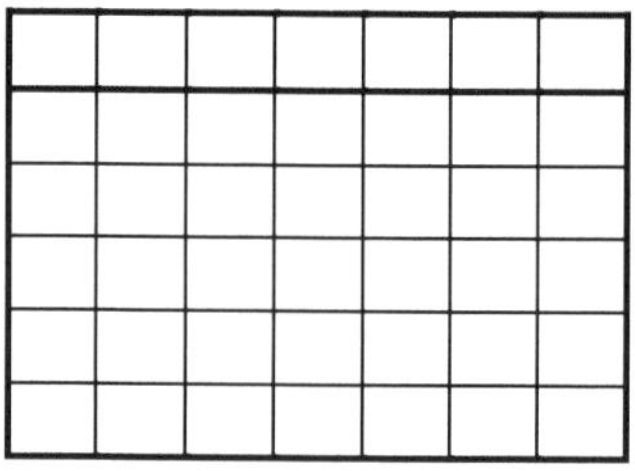

Figure 7.1 The school as modernist grid.

teachers and students working within the little slots of time and in their own little separate spaces. Each box represents a discrete time-space filled with a designated activity, each activity neatly bounded, and each bounded entity connected vertically, horizontally, and sequentially in a variety of predictable ways. As curriculum, much the same rationality operates. Each little box is bounded within a subject, strand, and level and each little box contains a little piece of knowledge that must be taught, measured, and recorded. There is room to move within each separate box, and sometimes even within aggregates of boxes, but the grid itself is hard to reach (Thomson 1999a).

The current UK version of the grid generally features forty-minute lessons and regulated movement between egg-crate rooms holding one teacher-one class (Tyack and Cuban 1995). There is very tight framing of particular knowledge into subjects which support and are supported by practices of tracking and setting (Gillborn and Youdell 2000). Adherence to this structure is maintained via the proliferation of managerial practice and performance audit and inspection (Gewirtz 2002).

Students who fail to fit into these time-space arrangements are not hard to spot. They can be seen in obvious places within the school and its surrounding neighbourhoods (see Figure 7.2 for a list of places a researcher might look to find them). Many stand out in a crowd, drawing attention with their choice of clothing, hairstyle, and shoes or their demeanour, stance, and expression (Eckert 1989; Thomson and Gunter in press).

Timetable: bottom sets; practical classes such as art, sport, technical studies; vocational courses, remedial classes, alternative programs.
School artefacts: list of absences and exclusions, bottom of test result sheets, whiteboard notices in staff room, daily notices, statistical returns to local authority office … student 'case' file particularly critical incident reports, referrals to behaviour support, correspondence with parents, monitoring sheets, psychological assessments
Places: back of room, in front of teacher's desk, outside classroom door, in the time-out room, counsellor's office, learning support centre, deputy or principal's office, toilets, far end of school yard, chip shop, supermarket, train station, high street.

Figure 7.2 Researching the spatialities of school exclusion.

Figure 7.3 Time-(out)-space.

Students who are candidates for school exclusion do not simply become visible by virtue of their appearance. They also make themselves obvious within the confines of the formal curriculum. They call out in class. They arrive at lessons late. They act out when challenged by teachers. They scrap in the schoolyard. They walk out of the deputy head's office rather than wait to receive their reprimand. They are urged to take subjects that interest them, that involve work 'with their hands'. They are highly visible in the school in both their presence, and in their absence. Staff (and some of their peers) know when they are not in class and breathe a sigh of relief.

Schools traditionally manage the behaviour of such students by manipulating both space and time. Students are sent out of learning time-spaces to punitive/therapeutic ones, such as the appropriately named time-out room (see Figure 7.3). They are required to attend detentions or to stay away from school time-space altogether for a fixed period. Sometimes they have alternative programmes and remedial classes created for them in which they leave the remainder of the student body to get on with the mandated work.

However, the perceived space in which students are required to fit does not simply map onto the lived space of students. In lived space excluded students experience themselves as visible to the school, but also paradoxically feel themselves to be unrecognized.

The lived space of exclusion

Students who challenge the social order in schools rapidly get a 'reputation'. While this might be enjoyable at the time, it also leads to their being obvious targets for disciplinary activity. Students who are excluded regularly describe occasions on which they are singled out for events which were the result of group action, rather than the result of their individual culpability, as in Julie's story below:

> I used to be dead cheeky and everything. Like every time a teacher walked past me who was a bit chubby I would say: 'Hey fatty bum-bum' and stuff like that. I did it because my sister says it to me. So I think it's alright so I say it to other people.
>
> I was shouting at the teachers and messing. I would never, like, throw a chair at the teacher. I would just, like, shout and walk out. 'Cos I never like it when a teacher shouts at me and every time they do I just shout at them back.
>
> Well then this girl set a fire and I was with her and there was loads more people and they blamed it on me. And before that they all picked on me. They blamed it on me because they know that I'm the 'darest' one—because we used to do a lot of dares. But I never would do fire.
>
> I'd been in trouble before loads of time. Being naughty and that. And all the others that were there were in trouble once or twice. They had fifteen days of exclusion and got put on report. They permanently excluded me. I thought: 'They are not my friends really because if they were real friends they wouldn't have said it were me'. (CATE 2005)

The literatures on school exclusion are replete with examples of students like Julie who feel highly visible but also at the same time as if they actually are *not* 'seen and heard' (see Batten and Russell 1995). In a research report tellingly entitled *Listen to me, I'm leaving* (Smyth et al. 2000), a group of Australian researchers painstakingly documented the experiences of 209 early school leavers. The young adults they interviewed were adamant that the vast majority of teachers had not stopped to ask what caused their problems at school. They had been treated as if they were simply a difficult anybody/nobody:

> basically treated like a non human being.... Every person is different and everyone has to be dealt with differently.... So if a kid's really stuffing up and doing some terrible things you need to get them and sit them down and say, you know, what's going on? What's your reasons for doing it? What do you want to do? Like where do you want to go in life? Because no one wants to be in trouble all the time. No-one wants it. (Smyth and Hattam 2004: 63)

In this particular study students were visible within their schools, but because of the ways in which they felt they were treated, they experienced neither being understood nor heard in ways that would help them to keep up with their education.

The key to understanding this apparent paradox of visibility and invisibility lies in conceived space and the specific exclusionary practices of representation.

Conceived space and the construction of borders

Disorderly and undisciplined students are made symbolically visible to staff, peers, and researchers through what Sibley (1995) calls the 'geographies of exclusion', the production of bordered spaces in which only some are included. Sibley argues that the powerful speak and act to create and maintain space which protects the borders of their 'self', they will eliminate, even persecute, anyone or anything which is 'polluting' or 'defiling' their psycho-social territory (1995: 10). Further, the powerful extend the boundaries of their 'self' to include the physical space that they occupy, a process in which the person and their space come to stand for each other. Removing the threat to the self entails excluding the person or object embodying 'impurity' from their material place (Douglas 1966). Sibley (1995) argues that this practice of self-space protection creates moral landscapes, inhabited by virtuous communities, dependent on 'the maintenance of order, conformity and homogeneity' (p. 38). This is accomplished through a 'language of defilement' (p. 55), a particular set of utterances and behaviours which create a 'generalized other' who is de-humanized and made abject (p. 9). Sibley calls this stereotyping, and notes that it is associated with fear and anxiety and is often married with metaphors of blackness, dirt, disease, nature, and savagery.

In schools, this creation of a moral landscape through the generalized othering of young people occurs through specific language games which are not overtly vicious or shocking. The educational language of defilement is not about profanity, but it is about the unambiguously 'bad' who either change, or are separated out from their conforming peers so that they do not pollute learning time-space. Students who are likely to be excluded are often described as psychologically damaged and deficient, unable or unwilling because of their class/race/locality to conform to the requirements of the school:

> they have low self esteem. These families are in general—I don't use the word troublesome families—but they are maybe large families, they have little discipline themselves at home. They may be latchkey children. There is a variety of different reasons and then, suddenly, they are going into adolescence. (Special Needs Teacher, CATE 2005)

The existence of this 'deficit discourse' among teachers is hardly new news (see Valencia 1997) and it is a stubborn feature of everyday practice in working class schools. The history of educating the poor is marked with attributions of lesser genetic capacity, cultures of poverty in families and neighbourhoods, and inadequate parenting (see Thomson 2002). But the conceived space through which excluded students are othered extends well beyond the school. Policy driven discourse designates such students as 'special needs', or 'emotionally and behaviourally disturbed', as Hamill and Boyd exemplify:

> defining social, emotional and behavioural difficulties is a complex matter and the definition will, to a great extent, depend on an individual's perception of what is *normal* and acceptable behaviour. (2002: 113, my emphasis)

Excluded students are represented as abnormal, difficult, and deviant. They are not 'pupils' but a series of normative and normalized constructions which are often highly medicalized or criminalized (see Fine 1995; Harwood 2005; Lloyd 2005; Rasmussen 2005). Their capacity to 'answer back' to these representations is limited (Smith 1993; Thomson 1999b). They have little choice to do anything other than accept or reject the identities written onto them.

Sibley argues that once the categories describing/inscribing 'the other' are established, 'normal' social/self space must be protected through the erection of boundaries, and this is effected via the structuring of social space. Interestingly, Sibley applies Bernstein's (e.g. 1973, 2000) work on families and schools to spatial boundary setting and maintenance. Bernstein developed the notions of *classification*, which Sibley defines as boundaries which separate and define subject areas, and *framing* which he understands as 'distinctions between what may and may not be transmitted between subjects' (1995: 80). In words that resonate with the grid (Figure 7.1), Sibley (1995) proposes that institutional social spaces can be 'strongly classified...where there is internal homogeneity and clear, strong boundaries...inflexible rules determining internal arrangements' (p. 80). In such situations 'difference in a strongly classified and strongly framed assemblage would be seen as deviance and a threat to the power structure' (p. 80).

Sibley argues that conceived space is translated into perceived space—it becomes material. While completely eliminating the educational threats produced by the disorderly other requires an exercise of 'positional power' (Bernstein 2000), everyday exclusionary practices are beneficial to, and condoned and exercised by everyone who is included in the moral geographies of the school. Border maintenance becomes a widespread social practice.

> If *those kids* are away from school, for whatever reason then the whole ethos and the ambience of the lesson changes. It's more positive and constructive and *pupils* will say to you, 'Please sir, why do we have to have them back in? It's brilliant when they are not here'. (Special Needs teacher, CATE 2005, my emphasis)

In reality, both teachers and pupils construct educational normative relations. Some pupils are very active in constructing a classroom where those who 'stand out' are literally 'eased out' of the space if they refuse to comply:

> The group actually take over and first of all try and change the behaviour of that pupil in the classroom. But if that doesn't work they try to isolate him and push him away from the group. (Deputy Head, CATE 2005)

Given that most pupils do adhere to the disciplinary requirements of school, there is generally little apparent incentive or reason for teachers to query the norms that constitute the geographies of exclusion. But there are some schools and teachers who do want to change the situation for pupils on the edge.

I now consider a set of possible options for avoiding exclusion. I construct some spatial possibilities for inclusion and take up Soja's exhortation, outlined at the outset of this chapter, that it is possible to reconstruct the lived experience of socio-spatial relations.

POSSIBILITIES: TROUBLING THE IN/VISIBLE GEOGRAPHIES OF EXCLUSION

Resistance and compliance

Students on the verge of permanent exclusion very often see their educational options as a stark choice of either complying or resisting. Some of those ascribed as outside educational boundaries are keen to challenge the normative status quo they experience. They refuse to stay 'inside' or 'outside' and can be 'seen' breaking out or in at regular intervals. Such resistance is an act of agency which removes the student from the situation in which they are denied a positive sense of self (Wexler, Crichlow, Kern and Martusewicz 1992):

> You wouldn't think he had much up top but by God he knew how to get out of or get into: break into, run away from, hide or whatever.... He'd have been brilliant in a prisoner of war camp. (Deputy Head, CATE 2005)

This spatial breaking out/in is analogous to de Certeau's (1988) 'tactics' in which agents who are unable to mount an effective opposition to a temporal-spatial 'strategy' which oppresses them can only take action in the moment. They can appropriate space in which they have freedom to do and be as they (apparently) please. But this is a temporary state. If resistant actions become ongoing they cease to be what de Certeau calls 'tactical' and they too become a 'strategy'. In the case of students facing exclusion from school a resistant strategy might be persistent truancy or escalation of the behaviour at issue. Such agency, as Willis (1977) pointed out in the case of the 'earoles', is most likely to lead to permanent removal from the potential benefits of education. In the case of CATE students and others like them, such strategies of resistance are likely to lead to unemployment and unemployability:

The other option for excluded students is conformity, the end to which punishment is directed. Students who are excluded often talk about the option of conformity—they consider what it might mean to be part of the crowd, to fit in, to do what is expected, to be a 'good' student. In a recent project that examined an intervention intended to prevent permanent exclusion (see the next section), conforming meant *getting your head down.*

> When I first come here I didn't know nobody and I just got me head down doing me work.
>
> I see a better perspective.... I just think I'm just going to get my head down in Year 9 because I've got my SATS and everything and then I've got work experience. So just get my head down.
>
> I wish I could put my head back down and do my work 'cos all I want to do is get another year back in school and try and do my work and get as many GCSEs as possible.
>
> Now I've got me head sorted I know that there'll be no messing about. Just get my head down and do my work. (Interviews with excluded students, CATE 2005)

Excluded students do not experience conformity as particularly agentic, even though of course they must make an active choice to conform. Students with their heads down become invisible. While they may no longer be blamed for things they did not do, the cost of invisibility is a continuation of the lack of recognition of their 'self' they experienced as non-conforming students.

But, as both Szudlarek (1993) and Bhaba (1994) point out, emulation of required behaviour is only one form of fitting in. The other option is simulation, a situation where the agent assumes the appearance of conformity but in actuality mocks those in power without their ever realizing it. Pretending to be a well-disciplined student, while mocking the teacher through satirical compliance, *is* an option used by some students. However, it is often beyond the realm of possibility for those students facing permanent exclusion who largely have long histories of anger and hurt with the school system, as Smyth and Hattam (2004) persuasively demonstrate.

So if student agency via resistance or compliance is an option with costs either way, perhaps teachers can provide less double-edged options through manipulating space.

Changing places

The Coalfields Alternatives to Exclusion (CATE) project operates among seven secondary schools and a range of service providers in a regional town recently made poor by the wholesale closure of mines. All of the CATE schools have a range of school-based programmes which support pupils who are failing to conform to norms of good behaviour and orderly conduct. These include counselling, provision of additional learning support, alternative curriculum programmes, withdrawal from classes on a part time basis, and short term fixed exclusions. A small minority of pupils however does not respond to this in-school support and the problematic behaviour continues. More intensive intervention is required in order to prevent such young people being permanently excluded from school altogether. Because external alternatives are less effective than students staying in mainstream education (see Munn, Lloyd and Cullen 2000), CATE supports students to move from one school to another for a 'fresh start'.

A CATE school is able to bring to the CATE management committee details of any student who appears likely to face permanent exclusion in the near future. The CATE committee is able to provide two kinds of support:

1. The student stays enrolled in their home school but CATE provides funding for one or more alternative programmes.
2. The student and their family are offered the opportunity to enrol in another school in the CATE cluster. If this offer is accepted, the new school may also receive additional funding for an alternative programme to enhance the likelihood of the new enrolment being successful.

While all CATE schools must adhere to the national curriculum and policy requirements, and while all of them serve communities whose life experiences have been fundamentally changed by the economic downturn, each school is a different and unique place (Thomson 2000). Through CATE helping them to change places, students are able to leave behind a set of problematic social relations to become embedded in different school practices and begin to form new identities and networks.

> I: What's good about this school?
> S: Everything. Just like, it's a different place. Like I haven't been to many different schools but I reckon this is the best school in (county). I'm going to get a better education here. (CATE student interview)

While CATE is not universally successful, it has helped the schools to significantly reduce the number of permanent exclusions (Thomson et al. 2005). Importantly, it has also helped a group of vulnerable young people. Away from a peer group associated with particular patterns of behaviour, and removed from situations where there are histories of poor communication and conflict, some pupils make the most of the chance to become somebody different.

> I had a boy who was permanently excluded for selling drugs at his last school and he's come in and he's been absolutely fantastic. He's come to me as a Year 8 and has fitted in very well. But we do have a philosophy at our school to always give the kids a second chance and we rarely exclude kids here. We accept quite a few children. (Deputy Head, CATE 2005).

> I get helped straight away unless the teacher is with someone else. (CATE student)

> When I first started going into lessons they gave me different work to everybody else that were a bit easier. So I'd be able to go straight through that and then they put me on harder work and I'd do that. And then they put me on a bit harder work until I'm the same as everyone in the class. (CATE student)

> I've been here for like five weeks every day. And I've got a good report…I'll probably spend more time in school than out of school. (CATE student)

Some parents saw the managed transfer as preferable to a permanent exclusion because of the way it avoided the stigma and a sense of rejection associated with permanent exclusion.

> You know what a stigma permanent exclusion is. A managed move is different. [Pupil] felt so upset by the fact that [old school] had permanently excluded him because of that sense of rejection. And when you do that to a kid who has so many other problems it's just another thing to heap on them because they are full of guilt anyway. (CATE parent)

All parents with students who had changed places highlighted the new sense of 'self' that had been produced in the new location.

> He didn't want to come here…but now he really loves this place and he's grateful for what they've done for him. Things have happened since he's been here that have made him feel so good about himself. (CATE parent).

> He gets so much praise for it and he's never had that before. (CATE parent).

CATE schools have worked hard to develop and implement the managed moves scheme. For it to be effective for students nearing permanent exclusion, teachers, and particularly senior managers, have to be committed to the principle of a fresh start, prepared to sort out difficult issues including confronting colleagues in their own and other schools, and being willing to take students in. What they can achieve through this strategy is a situation in which students are able to conform while building a new sense of efficacy and identity in their new place.

But there is another spatial option which can work for students on the edge of exclusion. It is one which focuses less on conforming, and much more on students' sense of agency and identity.

COUNTER-PUBLIC SPACES

The idea of a counter-public is explicated by Nancy Fraser (1993, 1995). Out of the very institutional practices that are exclusionary, Fraser suggests, social actors can collectively appropriate space through which they can oppose damaging stereotypes, and create alternative identities, narratives, and knowledge. This sentiment is in accord with Lefebvre (1991: 26) who suggested that while space serves as a 'tool of thought and action', as a 'means of production' and 'control, and hence domination of power', it also 'escapes in part from those who would make use of it'.

A counter-public space can be created through an alternative programme, a lunchtime activity, or a particular subject within the curriculum. It is important to stress that a counter-public is not the same as a 'safe space' to which students can retreat for respite or extra assistance or attention—although it will have that function. A counter-public space does much more than offer an escape from the oppressive grid of schooling and normative categorizations. A counter-public space uses the borders, rules, and norms of the school to create new boundaries around social-individual activity. Weis and Fine (2005) suggest that committed educators can create secure counter-public spaces where young people can 'challenge "common sense" (stereotypes) about themselves as well as others and engage intellectual and political projects that are, indeed, counter hegemonic' (p. 125).

Girls on the margins of schooling who were involved in an action research project aiming to inculcate citizenship behaviour were able to construct just such a counter-public space (Thomson, Harris et al. 2005; Thomson, McQuade et al. 2005). Teachers Vivienne and Kerry were keen to provide something for a group of junior high school girls and for other young people that would 'allow students to be energetic, proactive and positive rather than dwell on perceived problems with school operations' (Final Project report). They knew that some of the girls were interested in the environment.

> It all started when the State Environment Council held one of their workshops. Me and another friend of mine...we went to one of them and then we came back the next week and we decided to do something about the area we live in...like just help the place, like just to clean up and start some projects and get a group together for some help...and we went to Mrs McQuade and it went from there (B and B).

Vivienne and Kerry saw this interest in the environment as the positive focus they were seeking. The environment was something about which many students were concerned. Here was the opportunity to connect with youthful energies and passions in a project that the students would see as having a worthwhile purpose.

The two teachers applied for funding to support the formation of a new student group. Identified students were invited to a meeting which was also advertised around the middle school. 'There were 17 students from Years 8–11 involved, both male and female, with a range of interests and academic successes' (Final Project report). The funds which Vivienne and Kerry had won were to be the budget for the new organization.

The group decided to call themselves YEA—Youth Environmental Activists—and thus staked a claim to a collective political identity with resonance beyond the school. They designed a YEA logo for new school T-shirts which they were allowed to wear as an alternative to the regular school uniform. They established their goals as the production of a sustainability plan for the school and as a contribution to a better local/global environment. In addition to undertaking their separate projects, such as tree planting and litter clearing, YEA operated a grant programme. Students and classes across their school were able to apply to YEA for funds to undertake environmental projects.

> S 1: The application form...like they tell us how much they need...we discuss the stuff that they need and see if it can work and we discuss whether it is going to continue with what they're doing....
> S 2: ...We either give them the money or we sit down with them and talk it over and stuff, and then if it's good enough we give them the money and then we've got more projects.

YEA was also allocated a room in the middle school area. Having 'a room of one's own' was particularly important to the group of girls about whom Vivienne and Kerry were concerned. YEA met there during some lesson times and at break times and they were encouraged to decorate it.

> This is like our actual room...*to be known for something*, for helping something, and *be who you are*, because in class there's this whole lot of students...but in here it's just being yourself, and it's like your own little paradise and we're also doing projects and stuff (B and B).

YEA was important enough to be given an entire room to itself. The students in YEA were officially 'recognized', not as naughty or deficient, but as young people who were making a significant contribution to the school. The students clearly connected the material place, in which they felt in control and able to manage, with their emerging sense of identity, and autonomy. The room was an extension of their new collective persona and signified materially/spatially the change in their social position in the school. Similarly the T-shirts were also significant in producing a new visible and acceptable self.

> I: How come the T-shirts are so important?
> S1: Because they show everyone *who we are*, and if they've got any ideas *they know who we are*, so they can come to us.
> S2: And they can *recognize us* from a distance (B and B, my emphasis).

The students identified that it was important for them to 'get out' of the usual classroom space—'it gives us something to work on and it can keep us occupied and just keep us out of trouble...and it's fun, doing something that you love to do' (B and B). YEA gave them a strong sense of agency and a tangible shift in identity.

> ...having a good idea, having it listened to and acted on (S3).

> ...not having to go to (the principal) all the time, "Can we have $150 for this?" (M).

> We're just helping ourselves as well as where we live (B and B).

Through YEA the girls were also engaged in political activities outside of the school with both local and state government youth advisory bodies.

> Interestingly, while nothing else changed about their schooling, the girls began to have significantly less trouble in class. They were prepared to conform to the requirements of the (modernist) school without experiencing it as a loss of agency or identity.

> No-one controls us but we do the work we're supposed to do (B and B).

> It is the commitment to the environment that counts in YEA, not anything else. *It doesn't matter if you are a good or bad student in the rest of the school*, if you're committed about something then you can prove that you're good in here (M, my emphasis).

Vivienne, Kerry, and I have suggested that this project disrupted risky gendered geographies. Critical to this shift were: the opportunity to create a new and visible positive identity within the school, what the girls dubbed 'the tree-hugger'; building connections with the local and global environment movement; gaining visible representation on local authority and state advisory bodies through which the girls could have a voice and be heard; and occupying a changed and highly conspicuous place in the school via the provision of a room and a budget. In other words, the project allowed young women with no status to be seen and heard. We suggest that this amounts to the creation of a small 'counter-public' space.

REWORKING THE GEOGRAPHIES OF EXCLUSION: IN SUM

Investigating the geographies of school exclusion brings to the foreground the social construction of space and the spatial construction of the social. It also offers an insight into some possibilities for action—changing place and creating counter-publics—which begin to trouble the power-saturated and inequitable practices of schooling. Such options offer possibilities for young people to experience, through changed spatial relations, new opportunities for building identity and agency.

Of course, nothing in this chapter provides sure fire directions for transforming the perceived space of education made manifest as the modernist grid of schooling and the contradictory policies of standards, accountability, and inclusion. Nor does anything I have written disrupt to any great degree the production of conceived space of policy discourse in and through which particular groups of students are deemed ab-norm-al and deficient. That is a story for another time.

What I have focused on in this chapter is the lived space of the in/visible geographies of exclusion, the daily experiences of young people and the teachers who work with them. I have suggested that concentrating on and deconstructing these everyday temporal-spatial relations can provide some options for modest changes. These may be very small in comparison to the histories of the institutional (re)production of educational disadvantage, but they are potentially large in the lives of young people who deserve better from their schooling. As such they are worthy of serious consideration.

ACKNOWLEDGMENTS

I would like to acknowledge the contribution of CATE and YEA colleagues, Belinda Harris, Kerry Vincent, Richard Toalster, Vivienne McQuade, and Kerry Rochford, to the production of material I have reworked in this article. I thank them for permission to re-use the original data for my own purposes.

REFERENCES

Allen, J., Massey, D. and Cochrane, A. (2000) *Rethinking the region*, London: Routledge.

Armstrong, F. (2003) *Spaced out: policy, difference and the challenge of inclusive education*, Dordrecht: Kluwer Academic.

Batten, M. and Russell, J. (1995) *Students at risk: a review of Australian literature 1980–1994*, Melbourne: Australian Council for Educational Research.

Bernstein, B. (1973) *Class, codes and control*, London: Routledge and Kegan Paul.

——— (2000) *Pedagogy, symbolic control and identity*, London: Rowman and Littlefield.

Bhaba, H. (1994) *The location of culture*, London: Routledge.

Brodie, I. (2001) *Children's homes and school exclusion: redefining the problem*, London: Jessica Kingsley.

Campbell, C. and Sherington, G. (2006) *The comprehensive public high school: historical perspectives*, Basingstoke: Palgrave-Macmillan.

Cullingford, C. (1999) *The causes of exclusion: home, school and the development of young criminals*, London: Kogan Page.

de Certeau, M. (1988) *The practice of everyday life*, Los Angeles: University of California Press.

Douglas, M. (1966) *Purity and danger*, London: Routledge and Kegan Paul.

Eckert, P. (1989) *Jocks and burnouts: social categories and identity in the high school*, New York: Teachers College Press.

Fine, M. (1995) 'The politics of who's "at risk"', in B. Swadener and S. Lubeck (eds.), *Children and families 'at promise': deconstructing the discourse of risk*, Albany, NY: State University of New York Press.

Fine, M. and Weis, L. (1998) *The unknown city: the lives of poor and working class young adults*, Boston: Beacon Press.

Fraser, N. (1993) 'Rethinking the public sphere: a contribution to the critique of actually existing democracy', in B. Robbins (ed.), *The phantom public sphere*, Minneapolis: University of Minnesota Press.

——— (1995) 'Politics, culture and the public sphere: toward a postmodern conception', in L. Nicholson and S. Seidman (eds.), *Social postmodernism: beyond identity politics*, Cambridge: Cambridge University Press.

——— (1997) *Justice Interruptus: critical reflections on the 'postsocialist' condition*, London: Routledge.

Gewirtz, S. (2002) *The managerial school: post-welfarism and social justice in education*, London: Routledge.

Gillborn, D. and Youdell, D. (2000) *Rationing education: policy, practice, reform and equity*, Buckingham: Open University Press.

Gordon, T., Holland, J. and Lahelma, E. (2000) *Making spaces: citizenship and difference in schools*, London: Macmillan.

Griffths, M. (1998) *Educational research for social justice: getting off the fence*, Buckingham: Open University Press.

Hamill, P. and Boyd, B. (2002) 'Equality, fairness and rights—the young person's voice', *British Journal of Special Education*, 29.3: 111–117.

Harris, B., Vincent, K., Thomson, P. and Toalster, P. (in press) 'Does every child know they matter? Pupils' views of one alternative to exclusion', *Pastoral Care in Education*.

Harwood, V. (2005) *Diagnosing 'disorderly' children: a critique of behaviour disorder discourses*, London: Routledge.

Lankshear, C., Peters, M., and Knobel, M. (1996) 'Critical pedagogy and cyberspace', in H. Giroux, C. Lankshear, P. McLaren and M. Peters (eds.), *Counternarratives*, London: Routledge.

Leander, K. and Sheehy, M. (eds.) (2004) *Spatializing literacy research and practice*, New York: Peter Lang.

Lefebvre, H. (1991) *The production of space*, Oxford: Blackwell.

Lipman, P. (1998) *Race, class, and power in school restructuring*, New York: State University of New York Press.

Lloyd, G. (ed.) (2005) *Problem girls: understanding and supporting troubled and troublesome girls and young women*, London: Routledge.

Massey, D. (1994) *Space, place and gender*, Minneapolis: University of Minnesota Press.

——(2005) *For space*, London: Sage.

Moss, P. and Petrie, P. (2002) *From children's services to children's spaces: public policy, children and childhood*, London: RoutledgeFalmer.

Munn, P. and Lloyd, G. (2005) 'Exclusion and excluded pupils', *British Educational Research Journal*, 31.2: 205–221.

Munn, P., Lloyd, G. and Cullen, M. A. (2000) *Alternatives to exclusion from school*, London: Paul Chapman.

Nespor, J. (1997) *Tangled up in school: politics, space, bodies, and signs in the educational process*, Mahwah, NJ: Lawrence Erlbaum.

Oakes, J. (2005) *Keeping track: how schools structure inequality*, New Haven, CT: Yale University Press.

Osler, A. and Vincent, K. (2003) *Girls and exclusion*, London: Routledge.

Paechter, C., Edwards, R., Harrison, R. and Twining, P. (eds.) (2001) *Learning, space and identity*, London: Paul Chapman.

Parsons, C. (1999) *Education, exclusion and citizenship*, London: Routledge.

Pomeroy, E. (2000) *Experiencing exclusion*, Stoke on Trent: Trentham Books.

Rasmussen, M. L. (2005) *Becoming subjects: sexualities and secondary schooling*, London: Routledge.

Reid, A. (ed.) (1998) *Going public: education policy and public education in Australia*, Canberra: Australian Curriculum Studies Association.

Rubin, B. (2003) 'On different tracks: students living detracking reform at a diverse urban high school', in B. Rubin and E. Silva (eds.), *Critical voices in school reform*, New York: Routledge.

Sibley, D. (1995) *Geographies of exclusion: society and difference in the West*, London: Routledge.

Smith, D. (1993) *Texts, facts and femininity: exploring the relations of ruling*, London: Routledge.

Smyth, J. and Hattam, R. (2004) *Dropping out, drifting off, being excluded: becoming somebody without school*, New York: Peter Lang.

Smyth, J., Hattam, R., Edwards, J. Cannon, J., Wurst, S. and Wilson, N. (2000) *Listen to me, I'm leaving: early school leaving in South Australian secondary schools*, Adelaide: Department of Education, Training and Employment, Senior Secondary Assessment Board of South Australia/Flinders Institute for the Study of Teaching, Flinders University.

Soja, E. (1999) 'Thirdspace: expanding the scope of the geographical imagination', in D. Massey, J. Allen and P. Sarre (eds.), *Human geography today*, Cambridge: Polity Press.

—— (2004) 'Preface', in *Spatializing literacy research and practice*, in K. Leander and M. Sheehy (eds.), New York: Peter Lang.

Szkudlarek, T. (1993) *The problem of freedom in postmodern education*, Westport, CT: Bergin and Garvey.

Thomson, P. (1999a) 'How doing justice got boxed in: a cautionary tale for policy activists', in B. Johnson and A. Reid (eds.), *Contesting the curriculum*, Sydney: Social Science Press.

—— (1999b) 'How stopping it makes us go blind', *Education Links*, 59: 10–12.

——— (2000) 'Like schools, educational disadvantage and "thisness"', *Australian Educational Researcher*, 27.3: 151–166.

——— (2002) *Schooling the rustbelt kids: making the difference in changing times*, Sydney: Allen and Unwin.

Thomson, P. and Gunter, H. (in press) 'From "consulting pupils" to "pupils as researchers": a situated case narrative', *British Journal of Educational Research.*

Thomson, P., Harris, B., Vincent, K. and Toalster, R. (2005) *Evaluation of the Coalfields Alternatives To Exclusion (CATE) programme*, Nottingham: Centre for Research in Equity and Diversity in Education, University of Nottingham.

Thomson, P., McQuade, V. and Rochford, K. (2005) '"My little special house": re-forming the risky geographies of middle school girls at Clifftop College', in G. Lloyd (ed.), *Problem girls: understanding and supporting troubled and troublesome girls and young women*, London: RoutledgeFalmer.

——— (2005b) '"No-one's a good or bad student here": an active citizenship project as "doing justice"', *The International Journal of Learning*, Available at: http://ijl.cgpublisher.com/.

Tyack, D. and Cuban, L. (1995) *Tinkering toward utopia: a century of public school reform*, San Francisco: Jossey Bass.

Valencia, R. (ed.) (1997) *The evolution of deficit thinking: educational thought and practice*, London: Falmer.

Vulliamy, G. (2001) 'A sociology of school exclusions', *British Journal of Sociology of Education*, 22.1: 177–184.

Weis, L. and Fine, M. (2005) *Working method: research and social justice*, New York: Routledge.

Wexler, P., Crichlow, W., Kern, J. and Martusewicz, R. (1992) *Becoming somebody: toward a social psychology of school*, London: Falmer Press.

Willis, P. (1977) *Learning to labour: how working class kids get working class jobs*, Farnborough: Saxon House.

Wright, C., Weekes, D. and McGlaughlin, A. (2000) *'Race,' class and gender in exclusion from school*, London: Falmer.

8 Warehousing young people in urban Canadian schools

Gender, peer rivalry, and spatial containment

Jo-Anne Dillabough, Jacqueline Kennelly, and Eugenia Wang

Drawing upon the theoretical insights of cultural geography (e.g. Massey 1999) and the sociology of youth culture, this paper explores perceptions of 'peer rivalries'[1] and accounts of social exclusion on the part of economically disadvantaged male and female youth (aged 14–16) in one urban concentration in Southern Ontario, Canada. In particular, we examine the impact of contemporary urban youth class conflict and neo-liberal school cultures on the social formation of youth sub-cultural identities in the modern Canadian inner city. In so doing, we seek to assess the ways in which economically disadvantaged male and female youth perceive the influence of gender and urban schooling in shaping their conceptions of their social futures, which are viewed here, following Reay and Lucey (2003), as 'tied to the geography' of urban cities and school life.

Our overarching aim is to establish a preliminary hermeneutic and praxiological (see Mottier 2002) framework for understanding the 'formation of new youth subcultures which may function, in some degree, both as a response to, and a connection between, macro and micro forces of social change' (Gardner, Dillabough and McLeod 2004: 11). At the same time, we also wish to offer a phenomenological reading of youth sub-cultural identity formation and social exclusion which both accounts for youth culture in its contemporary expressive forms yet remains ultimately tied to a materially informed analysis which seeks to expose its stratified, historical, and symbolic character. An account of this kind seeks to expose how young people negotiate, in the world of the everyday, the varying degrees of alienation they experience and 'what they do with the cultural commodities they encounter' (Williams 1977: 17). In common with Goffman's (1959) concern with modes of representation in everyday culture, we too are interested in the ways in which the particular modes of representation young people construct for themselves, and for each other, constitute the ground for particularly powerful forms of sub-cultural identification, notably through the socio-cultural practices of peer rivalry.

Our starting point is to address the complex ways in which the differentiated effects of contemporary class conflict, cultural elements of social and educational change, and gender relations conjointly impact upon diverse young people, and are simultaneously negotiated by them, at the level of the local urban spaces and the educational institutions to which they find themselves tied. We do not, therefore, simply look to the concept of 'peer rivalry' as a property peculiar to working-class life, leading in the direction of a normative fate as 'criminals or deviants', and standing somehow outside the realm of political economies of hierarchy and stratification. Rather, we are interested in the representational modes of gender conflict that such rivalry takes and the manner in which cultural 'rules' or deeply sexualized 'territorial' discourses are manifest in the socio-spatial relations of peer conflict. We identify how gendered forms of 'symbolic domination' (see Bourdieu 1997)—often articulated by young people as conflicts in style and sub-cultural gender identity politics—are in the same moment embodied and utilized by young people to generate what Thornton (1995: 1) has named 'sub-cultural capital', embodied in 'the form of being in the know'.[2]

In our attempt to understand such representational modes of youth sub-cultural identification, the key argument we wish to advance has two aspects. First, drawing upon the results of an ethnographic study of urban youth sub-cultures in Canada, we argue that a broad constellation of competing processes are at work in re-contextualizing female and male youth subcultures. These processes are compounded and heightened in intensity as a consequence of young people's localised positioning within 'demonized schools' (Reay and Lucey 2003).[3] Examples of these social and cultural processes include: the re-organization of urban inner city space and schools by governments embracing elements of neo-liberal policy; the gendered impact of symbolic youth territory in 'demonized schools'; and the entrenched influence of historically shaped gender, race, and class relations. Such processes interact with young people's modes of representation in reconfiguring contemporary forms of 'subcultural capital' and youth subcultural identity in pursuit of social status in schools.

Secondly, we argue that contemporary youth subcultures are substantially founded upon the often covert relationship between global and local forms of symbolic domination operating beyond, as well as within, 'demonized schools'. The impact of powerful modernizing forces such as commodification has resulted in a profound social distancing of disadvantaged urban youth from established 'cultures of success' within a complex and stratified educational market. This tendency has been exacerbated by local forms of gender, race, and intra-class conflict in schools, as well as by the increased contemporary risks, contingent largely upon neo-liberal restructuring, constantly faced by such youth. We are therefore reluctant to assume that modernization in the forms articulated by scholars such as Ulrich Beck (1999) impact on all young people in the same way. Rather, we note that forms of historically informed domination continue to impinge strongly upon young people's sub-cultural identifications. The cultural form of such identifications, often played out as 'being in the know', and cultural conflict among young people who may share broadly similar class positions, reflect longstanding social relationships of domination in the state. This means that many of the cultural strategies young people draw on in their everyday lives resemble what Lois McNay (2000) has identified as 'deeply sedimented' forms of gender reproduction or, in Smith's (1990) terms 'objectified discourses' of the social order. Like other contemporary youth theorists, we too seek to suggest that theories of detraditionalization have exaggerated the degree to which gender norms have been reconfigured along with the so-called freedoms which such marginalized groups have come to attain (McLeod 2002). In this sense, 'being in the know' as a young person is best understood simultaneously as a social process (i.e. the acquisition of 'sub-cultural capital'), a spatial function of territory, and a cultural mode of representation which is embedded in the stratified and deeply sexualized practices of peer rivalry itself.

The chapter which follows has three parts. First, we present an account of our theoretical framework on youth sub-cultural identity, exploring the links between urban space, inner city schooling, and young people's 'phenomenologies of meaning' (Gardner et al. 2004), centring particularly upon youth accounts of 'peer rivalry' and social exclusion in schools. Secondly, we sketch the inner city contexts where young people's phenomenological accounts of their exclusion have been established and constructed. The final section draws upon interview data from the ethnographic study to further explore the formation of youth subcultural identity in the Canadian context.

THEORETICAL FRAMING

Our interdisciplinary theoretical approach addresses an interconnected concern with what Veronique Mottier (2002) refers to as praxiology, and what Goffman, Ricoeur, Willis, Williams, Cohen, and other earlier cultural theorists have variously identified as cultural phenomenology and phenomenological hermeneutics. These approaches are valuable for their illumination of urban youth sub-cultural narratives and the forms of meaning making that young people undertake in their everyday lives. In drawing upon these larger conceptual formulations, we attempt to chart new theoretical territory about youth exclusion which is not bound by familiar divisions between structuralist or culturalist debates which often position youth sub-cultural identity or youth identity at two polarized extremes. On the one hand, purely culturalist accounts of youth culture often find themselves tied to the significance of culture as locally derived sets of discourses which serve to regulate, rather than determine, young people's subjectivities and their commitments to particular youth sub-cultural identities. In this context, youth identity and subjectivity is seen as ever-changing, fluid, fragmented, unstable, and ultimately derivative of larger popular or state discourses divorced from material relations.

On the other hand, contemporary structuralist or high modernist positions suggest that youth identity is shaped largely by social structures, elements of social change, and global forces which are an outcome of what Ulrich Beck (1999) has termed 'reflexive modernization' (which challenge earlier neo-Marxist positions). In this case, youth identities are seen as shifting in line with advanced forms of modernization which are less concerned with entrenched class histories or unique local cultural communities. From this vantage point, wider global forms of class stratification and de-ritualized forms of change become the focal point of interest.

In refusing the straightforward dichotomies of competing structuralist or culturalist positions, we wish to contend that while greater levels of youth individualization represent visible and embodied markers of social change, inherited class positions and spatial forms of meaning making are simultaneously and powerfully at work in both youth identity formation and patterns of exclusion. At the same time, we argue that inherited regulatory mechanisms of deeply sedimented forms of masculine domination (see Bourdieu 1998) persist, continuing to shape the very conditions underlying the formation of youth sub-cultural identity. What this implies is that neither simply free-floating discourses, deeply entrenched inherited culturalist formations, nor modernizing global forces can on their own be seen as singularly responsible for the formation of youth sub-cultural identity. The relational negotiations achieved by youth, along with their associated phenomenologies of meaning, are rendered possible precisely through their daily exposure to complex cultural, spatial and historical milieus operating at the local, national and global levels. Young people's phenomenologies of meaning then, if always unique, are also always linked to the broader cultural operation of political and spatial power, social hierarchy and exclusionary privilege.

CONTRIBUTIONS FROM CULTURAL GEOGRAPHY TO ANALYSES OF YOUTH SUB-CULTURAL IDENTIFICATION AND EXCLUSION

Our theoretical position anticipates that youth accounts of exclusion and sub-cultural identification contain, within themselves, multiple layers of power relations, which operate as part of an unequal distribution of social space in the state.[4] As Massey (1995) makes clear, space constitutes an inherited dimension of social life, which is underpinned by historical and contemporary material realities that transform constantly. For Massey, space is, moreover, a site which differentially regulates economically disenfranchised male and female bodies and which therefore indicates that any geographical border within urban space 'is, in fact, very unevenly distributed' (Massey 1995: 337). In consequence, aspects of space are strongly tied to identity because they define who has the right to occupy a particular space or who may be seen to be 'out of place' (see also Wright 1997).

From this vantage point, we can link the formation of youth sub-cultural identities with 'micro-geographies of exclusion', 'territoriality', and urban containment in the cultural reproduction of stratification among youth. We can also recognize that any socio-cultural analysis of space—as a site of social struggle—must also account for a gendered ideology of power which conceals, in large part, its pervasive workings. Such forms of power, as represented in social life and in schools, may be seen as reproduced through culture and therefore manifested in the experiences of youth themselves, and their cultural understandings of past, present, and future.

CONTEXTUAL AND EPISTEMOLOGICAL CONTEXTS

We wish to highlight two contextual and epistemological contexts of meaning which can be derived from a spatial and cultural geographical analysis. We refer to contextual as the local contours of urban space, bringing together a particularly nuanced range of local urban configurations of inner city life. Epistemological refers to the ways in which these contexts are embedded in forms of meaning and significance for those young people who interpret them (see Gardner et al. 2004).

Spatial/Contextual

The spatial, as we have defined it here, is the multi-layered micro-contextual element of the urban space that youth inhabit and which plays a major role in the regulation and formation of youth identification practices. Such urban landscapes are not neutral. They are better seen as 'spaces of capital accumulation in highly urbanizing [...] arenas' (see Wright 1997: 5).

The urban inner city where the study took place has a long history of occupation by distinctive ethnic and class communities. When, from the late twentieth century, traditional industries began to lose their dominant place as a source of employment for many working-class communities[5] the former industrial area began to experience varying degrees of economic disadvantage and has, to a greater or lesser degree, remained so until the present day. Currently, the area is characterized by low cost tower apartments, together with housing which is self-owned, rented, or subsidized by the state. In the years between 1980 and 2003, a number of important urban transformations took place.

First, social support for families living in this part of the city was cut substantially forcing many with children to engage in full-time employment, as well as sometimes making use of their children's part-time income. Many parents of the young people in our study were working two to three contractually negotiated jobs each or were working in factories outside of the neighbourhood, usually due to the relocation or closure of businesses.

Secondly, the neighbourhood itself was shifting in accordance with recent zoning policies which allowed for the presence of transient forms of social housing, recently built low cost non-state housing, and state supported housing. While it is important not to pathologize a neighbourhood dedicated to public projects, urban re-zoning undoubtedly contributed to the loss of 'traditional local integration' (Cohen 1999) and to the re-shaping of the neighbourhood as a site of spatial containment. Such changes also meant that the idea of a neighbourhood of trust, if ever firmly established, was now under substantial threat in this new, leaner economic landscape. In the late 1990s rental prices did in fact temporarily rise, leading to a partial gentrification of the area.

At the time of data collection, there were signs of the emergence of an urban mix of those identifying themselves as lower middle-class (e.g. 'artists') and professionals (e.g. 'working-class teachers'), alongside underemployed families or low income families who lived in state-funded housing. Cumulatively, what is most significant about this area is that it was widely seen by many young people in the study as a site of abandonment by government, city officials and others.

National Context

In order to understand urban inner-city spaces in the national context, a number of cultural, economic, and educational shifts must be seen as significant for the lives of the youth participants. The gentrification of inner city centres shifted the location of urban economic disadvantage to neighbourhoods on the fringes of the city proper. At the same time, youth economic disadvantage rose across urban Canada from the 1980s (see Human Resources and Development Canada [HRDC] 1998), increasing by 46 percent in little more than a decade. Rising levels of youth economic disadvantage in these years were closely associated with the pronounced anti-welfare reform stance, legal and educational reforms (e.g. anti-squeegee laws, zero tolerance policies), and the spatial reorganization of urban cities and schools in many Canadian provinces (CCSD 2000). In ideological terms, such policies can be seen to draw powerfully upon global neo-liberal discourses and practices, operationalized primarily through retrenchment in welfare, public housing, education, and social service provision (HRDC 1999).

Schools located in the inner city have been substantially affected by such broader economic shifts, particularly in terms of educational choice policy and the publication of league tables[6] If schools appear as 'demonized' then many parents who live in surrounding urban areas of the school (e.g. 'regenerated' or mixed neighbourhoods) increasingly 'choose' to send their children to other 'higher performing' settings, thus creating both competition between schools across class communities—what some young people in the urban school context referred to as a 'ghetto' or 'warehousing effect'. This school stratification effect was compounded by larger global trends which were manifested locally in the school context. School success was rendered more difficult as the social conditions for learning had been undermined by program cuts which would traditionally have supported this group of young people. At the same time, the student population was becoming increasingly transient, with many young people obliged to move from public housing complexes that were overcrowded and, in many cases, also having to change schools. Students lived both with the burden of the stigmas attaching to 'failure' and with all the deleterious effects of urban transformation. While participants did not discuss these disadvantages using the terminology of neo-liberalism, it was eminently clear that they understood the role that reforms were playing in their lives. And there were times too when students also took the responsibility of blame upon themselves. The constant reminders, operating through heightened forms of individualization in educational rhetoric, that participants were not truly deserving students when compared to middle-class pupils, meant that the levels of fear associated with neo-liberal regimes of testing were particularly acutely felt.[7]

SPATIAL AND CULTURAL LANDSCAPES OF ETHNOGRAPHIC INQUIRY

In keeping with an ethnographic form of inquiry which draws photographic metaphors which might allow us to get 'close up' and 'personal', a combination of both spatial and cultural ethnographic techniques consonant with a phenomenological and praxiological approach were utilized. Over the course of one year, the first author engaged in both classroom observations and joint ethnographic projects with young people in an urban inner city school we call 'Tower Hill'. The aim in the methodological approach was to capture the phenomenological meanings generated about young people's everyday life experiences in 'demonized' schools, alongside accounts of their social experience of exclusions in urban space. We also wanted to understand better how young people come to understand sub-cultural capital, and how they draw upon it in their peer rivalries. In this way we hoped that we could say more about the meaning making practices and modes of representation of contemporary urban youth, as well as the larger forms of symbolic domination which continued to operate within the 'fields' or 'spaces' where young people played out their everyday lives.

In confronting elements of the praxiological, moving beyond a traditional interviewing format was crucial for understanding young people's cultural worlds. We drew broadly on a project-based approach which allowed young people to engage in a variety of activities across a period of one year, designed to capture in non-abstract ways, their diverse accounts of school inclusion/exclusions.[8] In classroom contexts, we asked students to complete activity-based project work involving, for example: developing time lines on the subject of exclusion, work, and social change; developing critical media/film accounts of urban cities; and completing urban photo-narratives (using disposable cameras) encapsulating young people's experience of social space. To contextualize these narratives, we gathered substantial qualitative data on the urban neighbourhood and its surrounding areas, including tracking zoning policies and urban transformations in the area over time. We also drew upon Luttrell's (2002) exemplary methodological arguments supporting a 'person-centred' research approach. This approach posits that obtaining experiential accounts of, for example, peer rivalry or objectified discourses of exclusion is in no way designed to further 'demonize' youth groups or render them 'strange'. Neither the research design nor the data it generates lead in any such direction. Rather, exposing young people's modes of representation served as the *hermeneutic imperative* (Gardner 2003) for understanding better how young people make sense of what Hannah Arendt (2003) has called the 'social life world' and its link to youth exclusion in the broadest of senses.

The lead author also engaged in individual open-ended interviews with the young people. Interview questions addressed students' accounts of their schooling and social experiences across aspects of urban space in times of change; their descriptions of school culture and urban space; their view of themselves as individuals and 'citizens' of Canada and in relation to peers, popular culture, and issues of cultural/national identity; and the impact of urban life upon their school experience. The interpretation of the data generated through interviewing was further developed by drawing from additional data sources such as field-notes, observational data, and documentary information relevant to the site in question.

The youth participants, ages 14 and 15, were located across a range of identity and cultural positions. The school drew both from 'migrant' and 'refugee' communities, as well as local white working-class and Italian and Portuguese communities. There was also a small minority of First Nations students who attended the school. Many of the youth participants were children of working parents, single parent families, and families who might be underemployed or classified as low-income working families.

YOUNG PEOPLE'S PHENOMENOLOGIES OF MEANING

Sexualized territory, peer rivalries, and the struggle to classify

In this section, we focus particularly upon gender and other elements of social location as key categories of analysis in exposing young people's 'phenomenologies of meaning', articulated through their narratives of peer rivalry and exclusion. We view these narratives as objectified discourses which are ideologically bound and located in the historical subordination of gender, culture, and particular forms of masculinity and femininity. The continuing symbolic power inscribed within these cultural narratives means that the 'subjective work' involved in meeting the expectations of an 'ideal of the perfect girl' (Reay and Lucey 2003) or 'perfect boy' combines the exercise of gender power, particularly in the form of heterosexual masculinities, and school experiences, with its standardized and artificially narrow definition of student success. The result is considerable anxiety as young people pursue deeply conflicted social and personal objectives.

In our study, larger discourses of the gender order, race relations and territoriality were often exploited by young people to gain *symbolic control* (see Cohen 1999) over both their educational and urban landscapes. We do not define territoriality here as an objective site which is claimed by young people but rather as a social process in which young people engage in what Bourdieu (1998) has identified as 'classification struggles' to determine their own social status (and acquire sub-cultural capital), along with that of others, within a peer group. Cohen (1999) writes:

> The function of territoriality is to provide a material basis for a system of positional rules which preserve the boundaries of the loose knit peer group network, and assigns the entire youthful population, big and little, boy and girl, to a place which cuts across these distinctions, a place which is marked by an unequivocal question: Friend or Foe. (Cohen 1999: 66)

Like Cohen, we show here that this seeking out of Friend or Foe—as central to the function of peer rivalry—is often one of the key sites for gender identification in schools, as well as serving as the basis for intra-class and racial conflict between young people pushed to the margins through class stratification. Peer rivalry, performed most visibly and tangibly in the school corridors, often served as the predominant site for the official and unofficial classification struggles that young people undertook in their efforts to reclaim meaningful symbolic territory in their lives. The corridor was therefore not just an oasis free from teachers. It was 'an emblematic class related place' (Massey 1999) where forms of social power and gender regulation could be exercised.[9]

What were the dominating phenomenologies of meaning emerging from the peer rivalries played out within these corridors of power? The most striking turned upon gender codes and objectified discourses relating to 'tough girl/boy' metaphors, as well as more traditional conceptions of femininity. The 'tough' girls/boys situated themselves in the representational mode of Gangsta culture (e.g. citing Rap and girl on girl bullying, boys' symbolic referencing was chiefly to baggy trousers, the wearing of chains and brand name clothing, and the carrying of cell phones). By contrast, the apparently 'soft, more geeky girls' who needed roughing up and sorting out were defined by the Gangsta Girls as the 'Ginas'. Ginas often objectified their cultural self-understandings as symbolically tied to contemporary dance music, snug clothing, and trying to complete homework.[10]

For Gangsta Girls, it was essential to conform—through *modes of representation*—to notions of the 'tough girl' whilst at the same time ensuring that particular notions of femininity remained intact. In this respect, the Gangsta Girls share much in common with Willis's (1977) and McRobbie's (1978) accounts of over-determined 'working-class' and gendered resistance to acceptable school norms of behaviour. Yet a new, more modern and assertive cultural account of female gender identity was enacted to challenge the officially sanctioned gender norms of the 'good girl' or the 'nice girl', which was deeply entrenched within local class and community cultures. At the same time, in displays of deeply paradoxical gender behaviour, some of the 'tough girls' (as well as Ginas) also encouraged the surveillance of traditional forms of femininity—girls as objects of consumption (e.g. demanding short skirts)—in conformity with established definitions of femininity and compulsory heterosexuality.

For those who practiced it, objectifying the ideological discourse of 'toughness' brought the advantages of a perceived form of agency and symbolic creativity (see Willis 1977, 1990) for girls who actively attempted to manage the conflicting relations of gender power in the school. Such an exercise was, however, always driven by ambivalence. Indeed, when such young women imagined themselves in the future, gendered modes of representations tied to their families, their working- class histories, and traditional forms of employment tended to dominate the consciousness of both the Ginas and the Gangstas. The cultural texts generated by the young women in our study speak powerfully to the active and symbolic creativity of constructing a future in a constrained social world. They also highlight how agency is bound by competing sets of gender codes which position women as workers of a particular nature on the one hand, and, on the other, as domestic servants or as agents of reproduction in the 'domestic register' of social life (see Cohen 1999: 368, see also Figure 8.1 below).

Let us now explore some of these emergent themes within young people's phenomenological meaning-making by looking in more detail at the case of a 14-year-old Portuguese girl, Lola, self-identifying as both working class and as a Gina. Lola reported being verbally and physically harassed by a group of Gangsta Girls who reprimanded her to the effect that her uniform skirt should be worn shorter.

> J: [you mentioned that] some kids [gangster girls] are bullies, do you think the majority?
> L: I think most are bullies.
> J: Have you had any experience with that?

Figure 8.1 A young person's vision of the future.

L: Well I have once cause I had my skirt too long, my kilt, and they're calling me [names], they would tease me well roll up your kilt or they'd push me around.
J: Ah-hah
L: So...
J: Physically push you?
L: Yeah
J: Girls or boys?
L: Some kind of girls/boys, but now I'm used to it. I just roll up my kilt now.
J: so they wanted it to be shorter.
L: Yeah.
J: Not longer? Right. Why? Why do you think?
L: Cause they thought I was geeky and nerdy.

How do we understand Lola's cultural experience as a working-class Portuguese girl, as well as the activities of her peers? One interpretation (beyond the more obvious forms of embedded gender domination at play) is that Lola represents a class foe, an expression of the primary class and cultural conflict between the Gangsta girls and the Ginas. We may understand this in Bourdieu's terms as a sexualized 'habitus', a heightened sense of approved gender behaviour which has evolved dialectically over time as a form of class conflict. Normative interpretations of gender by the Gangsta Girls were therefore fed by a form of spatial intra-class and gender conflict which served to define acceptable femininity in the school. This allowed the girls to define the culturally significant 'foe' and 'class enemy'—that is, to practice, enable, enforce, and simultaneously challenge certain forms of class stratification, cultural positioning, and gender division. Here, 'being in the know', possessing sub-cultural capital, meant learning—through *habitus formations* and '*horizons of possibility*' (Ball et al. 2000)—the correct mode of feminine representation in the classification of oneself as powerful.

Paradoxically, at the same time, masculine domination served as a form of 'invisible pedagogy' or set of restricted social codes (see Bernstein 1996) through which girls exercised their attempts to claim and redefine their class position. In this way, the forms of disciplining that took place served as an instrument of both male taxonomy and a representation of heightened intra-class conflict. One way of escaping the disciplinary process—of being exempted from being 'tough'—was to seek status through exploiting channels of personal consumption which connected with identities founded upon globalized cultural references. This strategy afforded some mediation of the class and race politics waged by other young women in the school context. The seizing of these global-local micro-channels of meaning offered some access to the very basis of acceptable gendered behaviour for girls. As Renissa, an Afro-Caribbean girl, reflected on the pressure to draw upon forms of the 'tough girl' to fit in:

> R At recess you'd always have to go and pick on a kid and make his life like hell.
> J: So do people put pressure on you to do that?
> R: Yeah, to fit in, to be cool but I never did it cause [*I had brand name so they left me alone*].

Another form of territorial and symbolic control was assumed by groups of girls who followed other girls around in the school corridors and outdoors at recess and lunch breaks. Much of this gender regulation served as the basis for positional competition over male 'boyfriends'. Often the direction of the conflict seemed manifested in the 'Gangsta Girls' understanding of themselves as 'being in the know' about correct sexual behaviour. This 'being in the know' functioned as a more severe form of gender disciplining when the targeted group of young women were racialized in particular ways (in this case, as 'Black'). As Wanda, an Afro-Caribbean student, reports:

> J: How did they get your number?
> W: I don't know, they'd just follow me [in the corridor] and they'd just start going, why mess with my man, bitch? Dah, dah, dah and I asked who's your man?...It's like shut up and leave me alone.

J: So you were upset, you had a boyfriend, what was going on there you think?
W: If it were true they would say the boyfriend's name and they would say the friend's name and they would know things about me but they don't know nothing. So they were trying to scare me [...].

Here is another account by Sarah, a second-generation Portuguese girl self-identifying as a member of the Ginas:

J: Nobody ever pushed you round or anything like that?
S: Yeah.
J: What happened, how did you handle it?
S: I would just be like don't touch me [] I wouldn't do anything. I'm not the type to swear or I'm not the type to be brave 'cause I don't wanna get my *ass kicked*.

In such cases, girls do not choose to be consumed by competition of this sort, or by the circulating discourse of *girl perfection*. Rather, they embody such ideals as part of a pre-existing gender order with very powerful local, cultural and spatial resonances. Within this order, such commonly used symbols as 'bitch' or 'messin round' speak to the inherited forms of masculine domination which frame working-class girls as essentially promiscuous. The 'foe', therefore, could be identified as in some degree outside the field of such resonances. In particular, the 'foe' could be associated with the racialized female immigrant who has stolen another young woman's man. When this happens such young women serve as 'scapegoat[s] of all the tensions circulating in the see saw of person to person perceptions' (Cohen 1999: 68). They ultimately become the target for the enhanced policing and exclusionary pressures which are associated with historically determined patterns of gender and colonial relations.

A further problem for some Ginas is that, whilst still identified as 'working class', they have somehow ended up on the wrong side of femininity when faced with gender competition from the 'Gangsta girls'. Here we see how McRobbie's (1978) conception of 'tough femininity' becomes a high status form of sub-cultural capital which represents a powerful way of 'being in the know' in public space. The mode of representation for these young women was, then, framed through mechanisms of compulsory heterosexuality. For example, young women often referred to young men as their closest allies in the struggle against the Ginas. The social texts circulating about 'being tough' and female–male partnerships thus served as the material premise which made the reproductive elements of the gender order possible.

But at the same time, such texts also subsumed these young women into a category of 'girl' which ultimately subordinated them to men. McNay's (2000) arguments about the durability of the gender order for particular groups of girls and women are very illuminating in this context. 'Gender identities', she writes, 'are not free floating: they involve deeply rooted investments on the part of individuals and historically sedimented practices which severely limit their transferability and transformability' (p. 12). Yet at the same time, understanding 'toughness' in the representational mode of the Gangsta Girl means understanding the gendered elements of change for girls themselves. With the shifting nature of gender as a social category, the discourse of toughness (when attributed to girls) was clearly under less official and moral scrutiny and played a more active role in shaping young girls' peer rivalries. In this way, we can begin to see the localised and temporalised significance of the 'Gangsta Girl'. She is configured as fundamentally new but also as locally familiar.

When we shift our gaze to the context of masculinity, we witness the deeply sedimented ways in which particular forms of femininity, especially when linked to modes of representation articulated by young males, are constructed as a form of class deviance. Gendered modes of representation therefore emerge as a set of cultural rules for determining one's place as insider or outsider or friend or foe. Here, Randy's words, a self-identified Thug and/or Gangsta, give some clues as to how gender emerges as the point of mediation between education and class:

> J: You don't like rock music?
> R: No
> J: Alright, what about the Gino stuff?
> R: No I hate that.
> J: What's wrong with the Ginos?
> R: I don't know, their clothes, their hair.
> J: Yeah, but what's wrong with it? [...]
> R: It's too girlish.
> J: So what's wrong with being girlish, Is there a problem with that?
> R: Well if you're a guy you can't be girlish, the school is filled with Ginos.
> J: Are they the ones that [you say] do well in school too, the Ginos?
> R: Yeah, they're the greasy people.
> J: But you know they might get a good job.
> R: It's not worth it for me [...]
> J: So, what if you drop out what are you going to do?
> R: I have no idea.
> J: Yeah, so what?
> R: Probably like every other Portuguese person, construction? Guys are just too cool for school.

The multi-layered expression of Randy's account is not only infused with gender domination. Rather what Randy is doing is accounting for, from his own experience, the role of territoriality and its associated rules for boys who are conscious of their marginal status. The 'rituals of resistance' and 'classification struggles' are therefore well established by a set of underlying phenomenologies of meaning which are not only shaped by history, but imply in the present that breaking masculine codes is not only a high risk activity; it is deeply humiliating. Randy's remarks can therefore be read through the corridors of power, but their real origins lie outside the school. The symbolic logic of 'becoming somebody', of being in the know, reflects a claim for sub-cultural capital through the assertion of masculine gender codes validated by the traditions of history and local space. Within the corridors of power the Gangsta or Thug is deeply committed—as a positional rule—with contesting the Gino mentality, perceived as a feminine artefact which is not only classed but constructed as a form of weak, diminished masculinity (Cohen 1999: 73). Randy's gendered account is also positional to the extent that its reveals the place of economically disadvantaged young males within the spatial division of labour. As Cohen argues:

> In working-class cultures, recognition is always of a difference, either one that is shared [i.e., gangsters] or one that isn't [ginos]. If such rituals involve a misrecognition, it is because these differences are regarded as contingent qualities emanating from—and hidden—inside individuals, rather than what they are: the structural properties of a system of relations governed by positional rules. It is a misrecognition of the grounds of identity. (Cohen 1999: 73)

As Cohen tells us, when economically disadvantaged young people seek out symbolic control through a 'classification struggle' they are attempting to master their environment, be it in school or in the city, in ways which are objectified through highly sexualized and material discourses. The embodiment of particular forms of masculinity and femininity therefore become important exercises in self-protection—in 'gaining closure such that a boundary can be created between those who are out to get you and those who are not' (Cohen 1999: 136). What we should note here, then, is that metaphorical references to 'Gangstas' and 'Ginos' link modes of self-representation with a form of self-protection, self-display, sub-cultural capital, and class reproduction which is consistent with the analyses which Paul Willis has so insightfully offered in the past. As Willis (2003) writes: 'although resistant cultures continue to be condemned in schools, they actually show some clear sign of rationality; resistant cultures supply shields from stigma to blunt the cruel edge of individualism and meritocracy in capitalist societies' (p. 394). Against this, it is by no means clear that Beck's notion of the de-ritualized self bears much relation to groups of young people, such as those attending Tower Hill, who live in difficult economic circumstances in an era of 'reflexive modernization'. For such groups, conceptions of gender symbolically organized within locally

Figure 8.2 A young person's vision of the future.

Figure 8.3 A young person's vision of the future.

defined histories and similarly constrained futures, remain very strong (see Figures 8.2 and 8.3).

PHENOMENOLOGIES OF MEANING ABOUT URBAN SPACE: 'AMBIVALENCE,' CONTAINMENT, AND SOCIAL CONSTRAINT

Some aspects of the complex geographies of exclusion lie outside the range of schooling or the more obvious forms of peer rivalry associated with the school. In this section, we turn away from the corridors of power in urban schools to the local contexts of urban neighbourhoods and urban spaces. Following Reay and Lucey (2003), we seek to explore young people's understandings of space 'as deeply ambivalent understandings of their social context'.

The social confinement of youth to particular places can be seen to contribute to what many youth referred to as 'warehousing kids' and what Bourdieu (1998) has referred to as being 'chained to a place' (see also Reay and Lucey 2003). This 'chaining' effect was often expressed by young people both as a form of ambivalence and of social constraint, as can be seen in the following interview with Clara, a 14-year-old girl who had recently arrived 'as refugee' from West Africa:

> J: [referring to photo narratives] What I would like about this place is it's very quiet and lovely. Is it very quiet?
> C: Yeah, it's very quiet.
> J: And no problems on the street or anything?
> C: No.
> J: So you live there with your family [sister]?
> C: Yeah. She doesn't have a job right now. My building someone got killed there. Yeah, in the elevator.
> J: Oh, what happened?
> C: I'm not really sure but I heard this guy got killed inside the elevator…I seen, yeah, and the guy got killed inside the elevator cause he…I think he was dealing with drugs or something.

The story which is hidden but immanent within this account is of the progressively worsening cuts in social support and public housing subsidies for economically disadvantaged young people from the early 1990s. Clara's construction of space was, in fact, intimately related to the temporal elements of social change and neo-liberal transformations of her urban neighbourhood. She must, however, in understanding her spatial location in the urban city, bear the 'weight of the world' as it manifests itself through working-class neighbourhoods in post-welfare economic regimes (Bourdieu et al. 1993). Her words convey how, as Massey (1994: 254) puts it, 'space is

Figure 8.4 Photo narrative A: Clara.

constituted through social relations and material social practices' (see also Clara's photo-narratives A and B below).

Although many young people expressed an interest in remaining in their neighbourhoods and schools, they were also all too aware of the constraints built into the web of systems in which their individual lives were embedded. As Cecilia remarks:

> J: So this is just where people hang and sometimes there are fights. Is this where you got into some of your fights, too, in that neighbourhood?
> C: Yup
> J: Yeah. Would you rather live some place else or are you happy there?
> C: I'm kind of happy but my mom just had a baby so I don't think it's the proper neighbourhood for the baby to grow up in. Right?
> J: Yeah? What are you gonna do?
> C: My mom's trying to move out of there (see also photo-narrative C and D below).

Young people mentioned strategies they drew upon—as objectified discourses—to keep themselves safe, such as saying 'hi' to an imaginary

Figure 8.5 Photo narrative B: Clara.

person in a doorway if they saw someone 'suspicious' nearby. Many also mentioned the regular drug use that they had witnessed on the streets and in the school. Some commented on the fact that these activities were taking place with impunity, as they witness no police presence to curb it. Their accounts open a window upon the 'landscapes of childhood' where negotiating place provides an understanding of the genesis of exclusion as lying outside the control of those who are objectified as both deviant and 'youth at risk' (Reay and Lucey 2003). They also give us some insight into a 'structure of feeling' founded upon fears of abandonment and exclusion (see Williams 1977; see also Reay and Lucey 2000) which appear grounded, in part, in the spatial and material conditions of the lives of young people at Tower Hill.

These structures of feeling are not located inside the marked bodies of economically disadvantage individuals who are sometimes seen to own such feelings as liberal subjects in the new 'global city'. Rather, their feelings of abandonment, containment, and ambivalence can be read as attempts to make sense of their lives in light of deeply contradictory social messages and experiences. On the one hand, young people are expected to absorb the dominant belief that their situations are simply theirs to make the best of 'we are not what we are but what we make of ourselves' (see Giddens

Figure 8.6 Photo narrative C and D: Cecilia.

1998: 34), and that they are limited only by their own 'skills' and 'willingness' to persevere. On the other, they daily confront circumstances which are beyond their control, largely the products of the neo-liberal restructuring of urban cities and schools marked and targeted as 'demonized'. This ambivalence often extended beyond students' urban location to their families' working situation where young people consistently affirmed that their situation was tolerable, was 'okay'.

There are myriad ways in which this 'structure of feeling'—as ambivalence in the face of adversity—has been theorised by other researchers. Reay and Lucey (2000) describe ambivalence as a form of working-class resilience, a manner of seeing things as bearable so as not to be overwhelmed. In bearing the weight of their worlds, Reay and Lucey argue that young people are engaging in the process of 're-classification'. 'Reclassification' is the action young people undertake to survive and make sense of the class conflicts between home, familiarity, and risk. Following

Wright (1997), another interpretation we might offer is that when a place (such as public housing) does not belong to you but rather you must belong to it, ambivalent readings should not be unexpected. Indeed, feelings of placelessness may be expected to dominate the consciousness of young people who do not choose a place called 'home' or a 'demonized school'. Clearly, most middle-class youth have greater spatial resources. They are often driven home in cars, live in property owned by their families, or walk home through leafy neighbourhoods, and have little cause to worry about the 'dangers' of urban inner city life. Such early degrees of spatial mobility reinforce social advantages accorded to those who live in privileged areas of the city. Against the knowledge of such mobility, the experience of economic disadvantage and 'failing schools' confirms economically disadvantaged young people's distinctiveness to others and, crucially, also to themselves as marginal non-citizens. In this way, place and youth sub-cultural identity are closely intertwined within the inner city. These elemental space-identity relationships are underwritten by a 'geography of social difference' (Massey 1994) which is rendered explicit through young peoples' phenomenologies of meaning. In short, young people's 'phenomenologies of meaning' highlight the power of urban geographies of exclusion in the very process of their operation.

CONCLUSIONS

Since the publication of Paul Willis's (1977) seminal work, *Learning to Labour*, sociologists of education have debated over whether schools are still profoundly implicated in the gendered reproduction of economic disadvantage. In contrast to Beck's individualization thesis, evidence from this study, from Tower Hill and its students, from the school corridor to urban space, suggests that economically disadvantaged youth are still profoundly tied to ritualistic, deeply 'sedimented' forms of gender/social relations, which are linked to class histories of work and education in the state.

Beyond these broad observations, the present research also highlights some key observations about peer rivalry in urban contexts as a site for gender and youth subcultural identification. First, we argue that intra-class conflicts and youth sub-cultural formations are linked to social location and geographical arrangements of power in urban schools and cities. Young people who are tied to particular geographical spaces and schools might therefore be seen as engaging in 'classification struggles' over the gendered meaning of youth sub-cultural identity. These classification struggles are an outcome of both inheriting and challenging, as young people, classed, gendered, cultural and spatial positions which are historically significant.

Secondly, we argue that any links made between neo-liberal school cultures, the reproduction of economic disadvantage and the spatial formation of social exclusion should reconsider the symbolic power of rules and gendered 'rituals of territoriality'. This return to a focus on rituals is significant as some contemporary structuralist accounts of youth identity imply that young people are less and less tied to ritualistic forms of cultural practice. Our data suggest that rituals still carry substantial symbolic meaning in the formation of youth sub-cultural identity formation. They appear to represent powerful forms of cultural regulation in the struggles young people undertake to gain social status in their respective youth communities. They also serve as an example of symbolic creativity expressed by young people who must navigate their lives in highly individualistic educational cultures.

Finally, we argue that symbolic domination, as a product of the cultural differentiation of discourse generated within urban spaces and schools, is at the very centre of male and female youth identity formation. Youth identities are therefore bound by cultural codes of masculine inheritance, serving as vehicles of gendered meaning, and as such, in Bourdieu's terms, are the embodiment of domination. Young people do not perform domination or objectify gender symbols as free, unfettered agents of the state but rather, as constrained social subjects, who attempt to resolve the contradictions which accompany the social circumstances of being contained in particular urban spaces and schools. At the same time, they are also actors who struggle daily to move beyond such embodiment. But one must be clear that the range of expressed cultural formations and choices available are substantially narrowed by political/symbolic economies of social exclusion operating on a far broader scale.

On theoretical grounds we therefore make the case, not unlike earlier youth cultural theorists, that culture and structure cannot be separated from each other in the study of contemporary youth culture. Discourse and materiality are not separate social entities. At the same time, the phenomenology of meaning making among young people is deeply cultural and symbolically located in the local formation of youth culture itself and impacts upon the material conditions underwriting young peoples' social circumstances and, very likely, their future destinies. In offering this account of youth sub-cultural identification, we do not assume therefore that young people are a product of their own making or that they are determined, as positivist sociological accounts might have us believe, by their 'class'. We need instead to view young people as responding and reacting to contested local histories confronting contemporary social change. This will allow us to understand young people's complex phenomenologies of meaning as emerging from a reconfigured history of marginalization 'set in motion' in the present.

ACKNOWLEDGMENTS

An earlier version of this paper was published in the *Journal of Curriculum Theorizing* (Summer 2005). We thank the editors of JCT for their permission to rework, and republish, this paper.

NOTES

1. See also O'Donnell and Sharpe (2004).
2. This evocative notion of 'being in the know' refers to the possession, by young people, of powerful insider cultural knowledge which serves at once to unify a particular subcultural community and to function as a divisive force within the broader social hierarchies of peer rivalry.
3. Following Reay and Lucey (2003), we define *demonized schools* as those sites which carry with them, as their institutional burden, a long history of 'under-performance' relative to middle-class schools, often maintaining low enrolments and which are under-resourced. Such sites are often threatened with closure or changes in administration as a result of their social/academic status.
4. This theoretical work was first developed in Dillabough and Van Der Meulen (under review) and has been revised for use herein.
5. We use the term 'working-class' in the way that British youth cultural theorists have to politicize the significance of young people's class history in shaping their lives in a stratified economic order. In so doing, we do not wish to undermine other forms of identification but rather to make the case that identification as a young person is deeply linked to class formations.
6. League tables refer to the ranking of schools by municipality on achievement scores emerging from both provincial testing and graduating averages.
7. Teachers also reported that funds designed to enhance literacy among various communities of youth had dissipated through program cuts. In this context, teachers themselves were being asked to do literacy training (many of whom had not been trained in ESL practices).
8. We completed these activities either alongside the classroom teacher, or in team teaching situations.
9. One 14-year-old Filipina girl illustrated this in remarking upon the language characteristically used to describe other teenage girls in the school corridor. The use of this language heightened in intensity when it was directed towards racially marginalized female youth: 'K: they'll [the girls] use this language... Slut...I don't know...Motherfucker...I don't know...'.
10. Noteworthy is that each of these classifications has a history in other forms of material culture in the UK, the US, and Canada. For example, the Thug or Gangsta fashion finds elements of its origins in loosely worn prison garb such as baggy trousers.

REFERENCES

Arendt, H. (2003) *Responsibility and judgement*, New York: Schocken Books.

Ball, S., McGuire, M. and McCrae, S. (2000) *Choice, pathways and transitions post-16: new youth, new economies in the global city*, London: RoutledgeFalmer.

Beck, U. (1999) *The risk society*, London: Sage.

Bernstein, B. (1996) *Pedogogy, symbolic control and identity*, London: Taylor and Francis.

Bourdieu, P. (1997) *Pascalian meditations*, Stanford, CA: Stanford University Press.

—— (1998) *Masculine domination*, Stanford, CA: Stanford University Press.

Bourdieu, P. et al. (1993) *The weight of the world: social suffering in contemporary society*, Stanford, CA: Stanford University Press.

Canadian Council on Social Development (CCSD, 2000) *Thinking ahead: trends affecting public education in the future*, accessed at http://www.ccsd.ca/pubs/gordon/part2.htm

Cohen, P. (1999) *Rethinking the youth question*, Durham, NC: Duke University Press.

Dillabough, J. and Van Der Meulen, A. (under review) 'A "magnified image" of female youth homelessness in Canada: history, society and the contemporary political subject'.

Gardner, P. (2003) 'Oral history in education: teacher's memory and teachers' history', *History of Education*, 23.2: 175–188.

Gardner, P., Dillabough, J. and McLeod, J. (2004) *Disenfranchised youth, class conflict and educational exclusion*, Social Sciences and Humanities Research Council Grant.

Giddens, A. (1998) *The Third Way*, Cambridge: Polity.

Goffman, I. (1959) *The presentation of the self in everyday life*, New York: Barnhardt.

Human Resources and Development Canada (1998; 1999) *Youth Report Series*, Available at: http//: www.hrdcdrhc.gc.ca/redirect_hr.html.

Luttrell, W. (2002) *Pregnant bodies, fertile minds*, New York: RoutledgeFalmer.

Massey, D. (1994) *Space, place and gender*, Cambridge: Polity.

——— (1995) *Spatial divisions of labour*, London: Macmillan.

——— (1999) 'Imagining globalization: power-geometries of time-space', in A. Brah, M. J. Hickman and M. Mac an Ghaill (eds.), *Global futures: migration, environment and globalization*, Basingstoke and London: Macmillan.

McLeod, J. (2002) 'Working out intimacy: young people and friendship in an age of reflexivity', *Discourse*, 23.2: 211–236.

McNay, L. (2000) *Gender and agency: reconfiguring the subject in feminist social theory*, Cambridge: Polity.

McRobbie, A. (1978) 'Working-class girls and the culture of femininity', *Women Take Issue*, London: Hutchinson.

Mottier, V. (2002) 'Masculine domination: gender and power in Bourdieu's writings', *Feminist Theory*, 3: 345–359.

O'Donnell, M. and Sharpe, S. (2004) 'The social construction of youthful masculinities: peer group cultures', in S. J. Ball (ed.), *The RoutledgeFalmer Reader in the Sociology of Education*, London: RoutledgeFalmer.

Reay, D. and Lucey, H. (2000) '"I don't really like it here but I don't want to be anywhere else": children and inner city council estates', *Antipode*, 32(4): 410–428.

——— (2003) 'The limits of "choice": children and inner city schooling', *Sociology*, 37.1: 121–142.

Smith, D. G. (1990) *The conceptual practices of power: a feminist sociology of knowledge*, Toronto: University of Toronto Press.

Thornton, S. (1995) 'Introduction to Part One: the 'Chicago School' and the sociological tradition', in K. Gelder and S. Thornton (eds.), *The subcultures reader*, London: Routledge.

Williams, R. (1977) *Marxism and literature*, Oxford: Oxford University Press.

Willis, P. (1977) *Learning to labour: how working-class kids get working-class jobs*, Farnborough, UK: Saxon House.

——— (1990) *Common culture: symbolic work at play in the everyday cultures of the young*, Buckingham: Open University Press.

9 Education and the spatialization of urban inequality

A case study of Chicago's Renaissance 2010

Pauline Lipman

> We're being pushed out of the city under the guise of school reform. (Teacher and community resident at a community hearing to save Englewood High School, a school serving low-income African American students in Chicago)

> As Latinos began to have more influence and power, it framed the debate differently, from 'Where are my children being bussed?' to 'Why don't I have [school] buildings in my neighbourhood?' (Ricardo Muñoz, Mexican-American city councilperson, explaining the 19-day hunger strike by Mexican-American mothers and grandmothers for a new high school in their community)

Chicago has been a forerunner of urban education reforms in the US over the past 18 years. In 1988, there was the creation of Local School Councils, perhaps the most far-reaching democratic local control of schools in the US. This was followed in 1995 by the takeover of the school system by the city's mayor, Richard M. Daley, and its recentralization under his appointed school officials. This new leadership inaugurated a high stakes accountability regime that was a prototype for George W. Bush's *No Child Left Behind* (NCLB) federal education law of 2002. Chicago's latest sweeping school reform, Renaissance 2010, announced by the mayor in June 2004, calls for closing 60 to 70 public schools and opening 100 new schools, two-thirds to be operated as charter or contract schools (a version of charters).

Historically, the US public education system has been decentralized. There is no national curriculum and elected local school boards make decisions about curriculum, text book selection, and assessments of student learning. Even so, the 1988 reform represented a further extension of 'democratic localism' in public education (Bryk, Sebring, Kerbow, Rollow and Easton 1998). The recentralization of the Chicago public schools in 1995 presaged a national trend toward greater centralization at local, state, and national levels with the implementation of local and state curriculum standards, high stakes assessments, and exit exams as a requirement for secondary school graduation.[1] NCLB introduced marketization and privatization nationally by requiring school districts to purchase supplemental education services from outside providers and through support for charter schools and the option for students in failing schools to choose new schools.

Renaissance 2010 (Ren 2010) is a local initiative that instantiates privatization and marketization in the Chicago public school system. The new charter and contract schools are a form of public–private partnership. Nonpublic entities such as for-profit companies, law firms, and community organizations receive public funding to operate the schools without public participation in school decisions. Teachers and other school employees in these schools are also not covered by city-wide school unions. At the Ren 2010 announcement, on 24 June, 2002, Andrew J. McKenna, Chairman of the Civic Committee of the Commercial Club of Chicago, an organization of the city's most powerful corporate, financial, and political leaders, said 'Chicago is taking the lead across the nation in remaking urban education. No other major city has launched such an ambitious public school choice agenda' (Civic Committee press release, June 24, 2004).

Two years before the announcement of Renaissance 2010, in May 2002, 14 people, mostly mothers and grandmothers, staged a 19-day hunger strike for a new high school in Little Village (La Villita), a large Mexican immigrant community in Chicago. The hunger strike was the culmination of a seven-year struggle for a new college preparatory high school. The community had one extremely overcrowded high school which was focused on trades. The only option for a college preparatory education was to gain admission to a selective public high school outside Little Village. Despite promises by school district leaders to meet the community's need, money budgeted for the new school in 1999–2000 was spent on new highly selective college preparatory magnet schools in gentrifying and upper income neighbourhoods of the city. It took the hunger strike to make the city fulfil its promise to the community. In Fall 2005, a $63 million state-of-the-art high school campus housing four separate college preparatory schools opened in Little Village.[2] The schools serve the Mexican-American community of Little Village and nearby African-American community of North Lawndale (more on this below).

Ren 2010 and the hunger strike represent competing interests and directions in education and community development in Chicago. They have unfolded in relation to global and national economic and social processes that have transformed Chicago from an industrial 'city of big shoulders', as poet Carl Sandburg famously labelled it, to a city of big finance. At the beginning of the twenty-first century, Chicago is home to corporate headquarters, global financial markets, business services, downtown development, and international tourism. Vast tracts of gentrification radiate from the city centre through neighbourhoods across the city. Glamour zones of culture, shops, and recreation for the affluent are coupled with disinvested and heavily policed low-income neighbourhoods of immigrants and people of colour.

In this chapter I draw on spatial theories to examine the relationship of education to political economy, culture, and race in the structuring of social life and relations of power in the city. I argue that education policy is implicated in the contest over spatialized inequalities, identity, and meaning in the city. I focus on the role of schools in the material and cultural reconstruction of urban space. I argue that education is central to struggles over capital accumulation, cultural appropriation, and domination in the city.

In addition to archival research, from July 2004 through to the present (February 2006), I have been a participant observer of numerous school board meetings, public hearings, picket lines, community and school meetings, rallies, press conferences, planning sessions, coalition meetings, and forums.[3] During this time, I have had ongoing conversations with parents, students, teachers, school level administrators, community organization leaders and members, members and heads of local school councils, the director and staff of a city-wide parent organization, congressional staff, representatives of teachers and school employee unions, school reform organizations, and community-based research groups.

Chicago school policy involves school closings and openings, transferring students to new locations, and claims to community and space—in short spatial change that is both material and symbolic. Education is bound up in struggles for housing, place, and community that are defining the spatial restructuring of the city—materially and culturally. To examine these relationships, I draw on the work of critical geographers, particularly scholarship that focuses on the spatial restructuring of the city as a feature of neoliberal policy (e.g. Brenner and Theodore 2002; Harvey 2001; N. Smith 1996, 2002). I also draw on perspectives from cultural geography to examine ways in which power is reproduced and contested, and daily life is regulated through the construction of meaning and identity in specific places (e.g. Keith and Pile 1993a, 1993b; Soja 1997, 1999). Bringing in these frameworks helps us think about the role of education in the class and racial restructuring of urban space. The centrality of racism in neoliberal policy in the US is often neglected (see Harvey 2005). As an enduring system of oppression rooted in the economic and political structures, ideology, and culture of the US, racism is central to the politics and cultural conflicts in the city. Racism plays a pivotal role in the accumulation of capital but is also a motivating factor in its own right as a structural and ideological system of maintaining racialized power and privilege (e.g. Haymes 1996; Lipsitz 1998).

THE POLITICAL ECONOMY OF URBAN NEOLIBERAL DEVELOPMENT

In part, the changing character of cities and metroregions—their significance in the global economy and their social differentiation—can be explained with reference to the spatial dynamics of capitalism. My analysis focuses on the economy and the role of capital in the production of neoliberal urban development, but I do not mean to suggest that transnational capital is the only actor in this process. (See M. P. Smith 2001, for a critique of economic determinist theories of globalization and urban development.) The transnational cultural practices and forms of resistance which also define the city are not merely superstructural aspects of the process of capital accumulation. They are constitutive of globalization as a contested sociocultural process which unfolds in locally specific ways in concrete spaces. Global economic restructuring opens up spaces for new social relations, solidarities, and cultural practices, new forms of human agency and resistance. This is clear in the opposition to Renaissance 2010 which brings together workers, professionals, community organizations, school reformers, and youth and in the Mexican solidarity and community consciousness that gave birth to the hunger strike for a new high school in La Villita.

Writing on space as a constituent element of capitalism, David Harvey (2001) argues that capital attempts to resolve structural crises through the 'spatial fix', the creative destruction and reconstruction of investment in the fixed environment. 'The inner contradictions of capitalism are expressed through the restless formation and re-formation of geographical landscapes' (p. 333). Or as Brenner and Theodore (2002) put it, the territorial organization of capital—the physical location of production facilities, the built environment of cities, places of consumption—is destroyed and rebuilt elsewhere in order to establish a 'new locational grid' (p. 355) for capital accumulation.

The reterritorialization of capital is always uneven. Some geographical areas are underdeveloped or disinvested in while others are over-developed, producing geographies of inequality, centrality and marginalization. 'Each round of capitalist development is associated with a distinctive, historically specific geographical landscape in which some places, territories, and scales are systematically privileged over and against others as sites for capital accumulation. The resultant patterns of core-periphery polarization and sociospatial inequality exist at all spatial scales' (Brenner and Theodore 2002: 355). Here I am particularly interested in the intersection of the global and the local in the uneven development of urban space.

Neoliberalism promotes the deregulation of national economies through transnational agreements such as the North American Free Trade Agreement and General Agreement on Trade and Services and extra-governmental bodies such as the World Trade Organization, International Monetary Fund, and World Bank. The policies of these extra-governmental structures' 'constitutional neoliberalism' (Gill 2003), have forced the rescaling of economic activity to local, regional, and transnational levels. Robertson and Dale (2003) point out that focusing on scale 'enables us to see that social activity is organized spatially, and that this spatial organization is itself highly political' (p. 6). As economic activity is rescaled, cities and metropolitan regions take on greater significance as economic actors in the global economy and as engines of capital accumulation. In this context, cities compete directly in the global economy. Interurban competition for investments, production facilities (including the production of business services central to globalization), and consumption represents an important aspect of global contention for economic power. This process is redefining the economies of cities and metroregions and their social geography, creating new economic, social, and spatial inequalities (see Lipman 2005, for a discussion of the Chicago metroregion). In major urban centres, such as New York, Los Angeles, and Chicago in the US, we see the contradictions of power and wealth that characterize the global economy as a whole.

GLOBAL CITIES

While some cities are marginalized in the global economy, others fight for position as niche cities, centres of specific production, finance, or tourism sectors. At the top of the urban hierarchy are 'global cities' which function as key actors in the global economy (Sassen 2005). Saskia Sassen (2005) argues that under present conditions of globalization, the simultaneous geographic dispersal of economic activity *and* its greater centralization in the hands of transnational financial and corporate interests creates highly complex coordination, financing, and decision-making tasks. She argues the complexity and interrelatedness of these tasks requires their concentration in specific locations—global cities—which function as command centres of the global economy. Global cities are marketplaces of global finance, strategic sites of innovation, coordination, and competition for the global production of goods, services, and financial speculation. Global cities are key nodes in an interconnected global network that represents a new scale of control and coordination of global economic activity (Sassen 2005).

Global cities disproportionately concentrate highly paid professionals and managers at the helm of globalization as well as low-paid service workers (primarily immigrants, women, and people of colour) essential to the functioning of business services. Thus, they are a concentrated expression of the contradictions of wealth and poverty that typify globalization (Sassen 2005). These contradictions are expressed in new forms of social segregation and dislocation and glaring disparities in the use of, and access to, urban space. Global cities are marked by upscale, gentrified neighbourhoods; corporate centres; and redeveloped downtowns catering to arts, tourism, and leisure on the one hand, and contained low-income African-American, Latino, and immigrant neighbourhoods on the other (Sassen 2005). The global city is a contradictory space of marginality and centrality, wealth and impoverishment, a spatially polarized landscape that epitomizes the core dynamics of power and inequality that typify globalization (Sassen 2002).

In 2002, Mayor Daley's office released its *Chicago Central Area Plan* for downtown development. The plan announced, 'This plan is driven by a vision of Chicago as a global city' (Central Area Plan 2002). Chicago meets many criteria of a global city. It has a high concentration of sophisticated producer and business services, international markets, and corporate headquarters (Abu-Lughod 1999). Its domination of the global market in futures and options makes it a 'first order international financial centre' (Reed in Betancur and Gills 2000: 27). Development of the downtown area, lakefront, and cultural venues has made Chicago a significant international tourist destination. Chicago is perhaps best characterized as a second-tier global city (behind paradigmatic global cities such as New York, London, and Tokyo) with first-tier ambitions (Moberg in press). A central aspect of that aspiration is the city's ability to offer the quality of life—housing, cultural and retail venues, and schools—that can attract the high-paid professionals central to key economic and financial functions in a global city. Chicago's downtown remake and gentrified neighbourhoods are located within this context.

GENTRIFICATION AS NEOLIBERAL URBAN STRATEGY

Neil Smith (2002) argues that gentrification has been generalized as a central force in neoliberal urban economic development. In the 1950s and into the 1970s, gentrification played a marginal role in urban economies. However, in the neoliberal context, gentrification has become 'a central motive force of urban economic expansion, a pivotal sector in the new urban economies' (N. Smith 2002: 447). 'Third wave gentrification' (Hackworth and N. Smith 2001), beginning in the 1990s, merges local, national, and transnational investment and speculative capital and is supported actively by the local state. Moving beyond the city centre, it transforms whole neighbourhoods into gentrification 'complexes' of consumption, recreation, culture, parks, and schools as well as housing. This clearly defines what has happened in Chicago.

Although it is cast as a positive strategy for urban decay and the achievement of social stability, present-day 'third wave' gentrification is a means for the middle- and upper-middle classes to claim political, economic, and cultural control of the city. Today's gentrification complexes represent the middle-class colonization of the city. The class nature of this process is, as Neil Smith points out, hidden in the language of 'mixed income communities' and 'regeneration'—or in Chicago's case 'renaissance'.

Ren 2010 is touted as part of a plan for mixed income communities with mixed income schools. By implication, there cannot be good working-class schools and communities—interestingly, those advocating mixed income solutions do not advocate class and race integration of white upper-income communities (see also N. Smith 2002).

Chicago's drive to become a first tier global city and the central role of gentrification in capital accumulation, have together redefined the social geography of the city. Chicago is characterized by affluent gated communities, refurbished greystone mansions, and blocks of luxury condos, upscale shops and restaurants—glamour zones for the affluent—alongside densely populated low-income immigrant neighbourhoods, and very low-income, economically disinvested African-American communities. Many working-class and low-income residents have been displaced as their neighbourhoods have become upscale residential and shopping sites.

The most dramatic case of this transformation is the displacement of thousands of public housing residents. Beginning in 1999, the Chicago Housing Authority (CHA) announced a plan to demolish over 19,000 units of public housing. It promised to replace them with new or rehabbed units in 'mixed income' developments (Chicago Housing Authority 2003). At its peak Chicago had 40,000 units of public housing, most of which were in huge high-rise buildings concentrated in racially segregated African-American communities. Over the past 25 years, the CHA failed to maintain the buildings, leaving them to deteriorate until many apartments were uninhabitable (Popkin, Cunningham and Woodley 2003), justifying their demolition. About half were officially vacant, although there were many unofficial renters living in these units 'off the books'. At the same time as the state disinvested in the buildings, working-class African Americans who lived there were hit hard by deindustrialization, cuts in social welfare, and disinvestment in community infrastructure. As a result, areas with large concentrations of public housing became some of the most impoverished census tracts in the US.

Sudhir Venkatesh and his colleagues (Venkatesh et al. 2004) who have been tracking the experiences of the residents of demolished public housing have found they have primarily relocated to racially segregated poverty areas. They have had inadequate counselling and social services, and it is estimated that about half of the squatters, mainly African-American men, were made homeless by the demolitions. Despite promises of new or refurbished replacement housing, Venkatsh et al. estimate only about 20 percent of former residents will be able to return to their old neighbourhoods, and housing authorities admit they don't have the funding to replace all the units (Public testimony, CHA official, Northwestern University conference, 'Gatreax at 40', 3 March 2006) nor do many former residents qualify.[4]

The combined demolition of public housing, the boom in expensive condominium developments, gentrification of older housing, and increases in property taxes have priced out renters and working-class home owners. The result is a crisis in affordable housing that is driving low-income and working-class people to the fringes of the city and beyond. A 1999 study found that 245,000 low-income renters in the Chicago area competed for 115,000 units of affordable housing (Catholic Charities 1999), and this was before the demolition of much of public housing. In 'hot' real-estate areas, housing prices increased from 25 to 40 percent from 1993 to 1995 alone (Schmid 1998).

At the same time, the downtown area is booming with new corporate and residential hi-rises, luxury condominiums, refurbished parks, museums, tourist attractions, upscale shops, world class restaurants and cultural venues. The crowning touch was the opening of Millennium Park, a new 24.acre, half-billion dollar public–private venture to create a 'world class' park, sculpture garden, and performance space on Chicago's downtown lakeshore. Thus, gentrification as capital accumulation strategy and the global city agenda are altering the city landscape and redefining the city's identity and who has access to urban space. As a whole, processes of disinvestment, displacement, and reinvestment are producing intensified race and class inequalities and spatially ordered social distinctions. Education policy is both motivated by and implicated in these processes.

RENAISSANCE 2010 AS NEOLIBERAL URBAN STRATEGY

Disinvestment in schools is an aspect of generating a 'rent gap' in a community and priming it for gentrification. Schools in Chicago's African-American communities have a legacy of inadequate resources (Orfield 1990), poorly qualified teachers, and under the district's recentralization, a series of changing central office mandates (Lipman 2004). A Midsouth community organization position paper asked 'Where is CPS Accountability for decades of second-class education? Where is the accountability for a probation system that has not worked, student retention (another failed initiative), direct instruction (a disaster), and social promotion. How many *thousands* of students have dropped out as a result of being mis-educated?' (Midsouth, original emphasis). The historical and persistent failure to provide adequate human and material resources and investment in schools in the most impoverished communities of colour leads to their academic and cultural devaluation to the point that there is justification to close them.[5] The schools are then reinvested in, revalued, and reopened as new schools to attract a middle-class clientele. In this way, education policy concretely contributes to the displacement of low-income urban communities of colour and the reconstitution of sections of the city as white and middle class. This trend is reflected in increases in the numbers of people living in poverty in suburbs just outside the city due to lack of affordable housing and the relocation of public housing residents (Mendell and Little 2006).

Gentrification as urban strategy is facilitated and managed by the institutional fusion of state and corporate power wielded through public–private partnerships and policies that use public funds for private development (see N. Smith 2002). The state uses its policing and administrative power to enact zoning ordinances that favour developers, raze public housing to open up new opportunities for speculative investment, step-up policing of low-income people of colour, and build new police stations, libraries, schools, and parks in areas being gentrified. In Chicago, a pillar of this strategy is Tax Increment Finance Zones (TIFs) that provide a public subsidy to developers by siphoning off tax revenues to support infrastructure improvements for real estate development. Privatization of public space is another aspect, exemplified by Millennium Park, which is closed to the public for corporate events. Ren 2010 represents yet another pillar in this strategy. Ren 2010 was proposed by the Civic Committee of the Commercial Club of Chicago, an organization that is actively and publicly engaged in social and economic policy in Chicago. The Commercial Club's 2003 policy document, *Left Behind*, calls for the 'creation of at least 100 public charter schools that increase parental choice and put meaningful competitive pressure on chronically failing neighbourhood schools'. The Commercial Club also promised $50 million to help fund the plan and created New Schools for Chicago (NSC) to coordinate its involvement. NSC, composed of corporate and financial CEOs and top school district leaders, selects Ren 2010 school operators to evaluate the schools' progress. This arrangement gives corporate and financial leaders explicit authority to make decisions about Chicago public schools. The direct involvement of the Commercial Club in framing Ren 2010 and funding and directing it through NSC, coupled with the Mayor's leadership in promoting the plan, instantiate the institutionalized fusion of corporate and state power.

The Commercial Club's close involvement indicates the strategic role of education in the neoliberal urban agenda. Global competition drives cities to engage in aggressive 'space marketing' (Harvey 2001; see also Brenner and Theodore 2002). Chicago's transformed landscape of lakeshore recreation and cultural venues, hotel complexes, new boulevards and landscaping of former commercial thoroughfares, and city-sponsored concerts and ethnic festivals are all part of marketing the city. New 'innovative' schools and choice within the public school system are also located in this dynamic. 'Good' schools and 'options' within the public school system are important in the global competition to attract investors and highly skilled knowledge workers. Good schools are also essential to attract investors to potential sites of gentrification and to subsequently market gentrified and gentrifying areas to new middle-class residents (Brenner and Theodore 2002; Harvey 2001; N. Smith 2002).

SCHOOL POLICY AND THE PRODUCTION OF SPATIAL INEQUALITY

Beginning in 1999, Chicago built six new selective college preparatory high schools at a cost of $30 to $50 million each. Five are in areas that are gentrified, gentrifying, or ripe for gentrification. The schools have state-of-the art facilities, intellectually rigorous curricula, and specialized courses of study in a college campus environment. These new schools provide an incentive for professional and middle-class families to live in the city, especially in areas of budding gentrification where they provide access to a *separate* high-status-knowledge program (see Lipman 2004, for a discussion). Ren 2010 extends the connection between school policy and gentrification. It is an example of the role of education policy in the neoliberal reconstitution of urban space. In the Chicago case, this is a highly racialized material and cultural process. I illustrate it with a discussion of three African-American communities that have been the focus of Ren 2010 in its first two years. In each case Ren 2010 is implicated in gentrification and displacement.

The 'Midsouth Plan'

Ren 2010 unfolded first in the Midsouth area, the most advanced example of the relationship of the education plan to the spatial reconstruction of the city. The Midsouth, an African-American community on the South side of the city, encompasses Chicago's historic 'Black Metropolis', of African-American institutions, arts, culture, and intellectual thought. This was also the location of two of the largest public housing complexes in the city, Robert Taylor Homes and Stateway Gardens. Robert Taylor alone consisted of two miles of high rise buildings—28 in all, containing 4,321 public housing units (Michaels). Three decades of public disinvestment, deindustrialization, and devaluation of public housing left the area one of the economically poorest in the US.

But the Midsouth has become a valuable piece of land and an emblem of the economic, political, and cultural processes displacing African Americans and gentrifying their communities. It has the advantage of location—easy access to downtown and proximity to expressways and public transportation, Lake Michigan, universities, and major medical centres—and a huge rent gap—the devaluation of the land to the point where it becomes a profitable site for reinvestment. The area is a bargain because the housing has been extremely devalued and huge swaths of land have been opened up by the demolition of public housing (Abu Lughod 2005). It is one of the most intense areas of gentrification in the city.[6] According to the Chicago Association of Realtors, the average sale price of a home in the Oakland community in the Midsouth rose from $76,000 in 2000 to $237,500 in 2004 and property taxes in the Midsouth area more than doubled between 2000 and 2003, driving out homeowners on fixed low incomes (Life 2005). Michaels Development Company has a $593 million investment in Legends South (Michaels), a complex of over 2,300 houses and apartments on a two mile stretch of land where Robert Taylor Homes stood.

Legends South is a paradigmatic public–private neoliberal venture—77 percent of the financing of the privately owned development ($407 million) will come from public sources (Michaels). The attraction of the area is increased by relatively lower prices for comparable housing available in other gentrified neighbourhoods. This marketing advantage is made possible by public subsidies. Developers got the land free under a 99-year lease from CHA and received a $4.5 million tax increment financing subsidy (Handley 2004: 3A). Michaels is just one investor in the area's redevelopment.

Although publicized as mixed income housing, much of it will be off limits to working-class and low-income families. Legends South, a 'mixed income' development, will provide public housing for just 18.3 percent of the former Robert Taylor residents (Venkatesh et al. 2004). Based on an estimated 2,700 Robert Taylor residents, over 2,200 will not have housing in Legends. Although African-American middle-class homeowners are moving in, the 'turnover' in the neighbourhood means it is whitening as well. In 2000, African Americans were 89 percent of the population and whites 6 percent; by 2003 whites were 34 percent of the homeowners (Abu-Lughod 2005: 300).

It is significant that the first step of Ren 2010 was the *Midsouth Plan*, made public in July 2004 (Dell'Angela and Washburn 2004). The Plan proposed to close 20 of 22 schools in the Midsouth area which was experiencing high levels of gentrification. Opposition from community residents, families, and teachers was immediate. At public hearings, meetings of the Board of Education, press conferences, and picket lines they argued that closing schools and transferring students to schools outside their community would push families to move out of the neighbourhood and lead to further gentrification. Teachers, students, families, and community members testified to the disruption and hardship the plan would produce—including lack of transportation to the new schools, fears for children's safety as they travelled across distances and often hostile territory to a different school, fragmentation of the school-community, and negative academic consequences of student mobility.

The Midsouth Plan was a school choice plan that seemed designed to appeal to the middle class. Students would no longer be assigned to a school based on where they lived. Instead, the region was to be divided into 'expanded attendance areas', each of which would include three to five 'innovative' school choices within walking or commuting distance (Bluestein 2005: 49). Over 90 percent of the students who currently attend Midsouth schools are from low-income, African-American families. However, the Midsouth plan stated the schools would serve 1/3 middle-income, 1/3 moderate-income, and 1/3 low-income students, and there was no guarantee that community residents would be able to attend specific schools. A Midsouth community organization said pointedly, 'What happens to the other 2/3 low-income students? Displacement' (Midsouth n.d.). With condominiums and townhouses in the Midsouth ranging from $175,000 for a one bedroom apartment to $500,000 homes in the Jazz on the Boulevard development (Geroulis 2003; Thrush), transformation of public schools in the area is a crucial inducement for potential buyers.

Englewood

In February 2005, CPS announced it was closing two elementary schools and a high school in Englewood, a very low-income African-American community on Chicago's South Side. Englewood is one of the most economically devastated areas of the city where there is not even a grocery store. Yet Englewood is located near to the expressway and public transportation and is discussed in Real Estate sections of local newspapers as perhaps the next 'hot' neighbourhood. A city Housing Department official noted, 'There is a growing interest in Englewood these days' (Almada 2005: 1, 16A). Without any new housing in the area in years, Englewood, a community with a median household income of $18,955, is now the site of a new $150 million development to build 550 new single family homes costing $165,000 to $365,000 and a second development of 185 houses. The neighbourhood revitalization plan is comprehensive: a $250 million city investment that includes a new police station and library; a new $150 million community college campus, and a new $22 million shopping district. At the same time, long-time neighbourhood residents are being driven out by up to 80 percent increases in property taxes and a 41 percent increase in house foreclosures in the past 10 years (Olivo 2004: 1).

A dominant theme in meetings of community residents is that closing Englewood High School and two elementary schools to reopen them as new schools is tied to gentrification. At a community meeting on the Ren 2010 school closings, a long-time community resident and veteran teacher at Englewood High School summarized many of the testimonies, 'We're being pushed out of the neighbourhood under the guise of school reform'. Like the community itself, for years schools have been floundering without adequate material and intellectual resources. Teachers at Bunche elementary school, one of the schools slated to close, reported they lacked up-to-date books, supplies, and professional development while they coped with an ever-changing series of external mandates and imposed programs (Teacher interview, 9-05).

North Lawndale

A third focus of Ren 2010 is North Lawndale, another extremely low-income African-American community on the city's West Side. In January 2006, CPS announced it was closing Collins High School and Frazier Elementary School over protests of parents, students, and community residents. Once again, the community saw gentrification written on the plan. Decades of disinvestment and declining population led to significant decline in the economic state of the neighbourhood, but it is located next to two expressways and major medical centres, is 10 minutes from downtown, and contains two extensive refurbished parks. North Lawndale has become a new 'hot' neighbourhood with a new 1,200 unit residential condominium conversion and new homes development underway. This development is next to a 300 unit residential development and there are other developments in the area. There is also a new shopping centre and a $47 million film centre. 'Evidence of investment is everywhere', noted the president of Royal Imperial Group, a Chicago-based development company. The area's city council person claimed the area has had more construction projects in the last few years than any other ward in the city (Almada, 4 September 2005).

At a meeting of over 200 angry parents, community residents, and students at Collins High School, speakers questioned why schools in African-American areas facing gentrification were the ones being closed. They claimed schools such as Collins that served some of the most impoverished families in the city had been set up for failure through school district policies that failed to provide the intensive support they needed. Problems were exacerbated in 2004–2005 when Collins was the recipient of students transferred from other failing schools. Collins is located in recently refurbished Douglas Park, one of the city's landmark parks. Around the corner are new $450,000 houses for sale. Collins is located just a half mile from developers MCL Companies and Brownstone Construction planned development of houses to sell for $250–600,000.

PATHOLOGIZING AND ELIMINATING BLACK URBAN SPACE

Space is both a material source of capital accumulation and a cultural context, invested with socially produced meanings. 'Space is both the "perceived space" of material spatial practices and the "conceived space" of symbolic representations and epistemologies. Every lived space is simultaneously lived and imagined' (Soja 1999: 74). Space and place are infused with socially constructed race, class, and gendered meanings that are produced internally by the people who live there through their distinctive spatial cultural practices and externally by media, government officials, and policed boundaries of inclusion and exclusion (Perez 2002).

Discussing the significance of Black urban space, Stephen Haymes (1995) argues that, while white supremacist ideology constructs Black urban communities as places of pathology and danger, African Americans, faced with specialized forms of regulation and control (i.e. urban racial segregation and white racist violence), have appropriated and defined Black urban spaces 'to renegotiate an oppositional identity' (p. 213). In Black urban communities, place making, and therefore the production of Black public spaces, is linked with day-to-day survival through collective support, affirmation of one's humanity, and resistance in contexts of racial terror. Thus, the dispossession of African-American working-class communities erases historically constituted spaces of collective identity and cultural resistance.

The representation of gentrification as benevolent rests on depicting the city as 'urban frontier' and areas to be gentrified as needing to be tamed (N. Smith 2002). Gentrifiers are 'pioneers', 'homesteaders', and the people they invade are 'savages'. Specifically, the path to gentrifying African-American communities is paved by their construction in media and government discourse as spaces of pathology. This discourse is rooted in a historically persistent white supremacist urban mythology 'that has identified Blacks with disorder and danger in the city' (Haymes 1996: 4). A recent Chicago Tribune magazine article on the 'rebirthing' of the Midsouth describes it as a 'neighbourhood that came back from the grave.... As drug trafficking, street crimes, and other negatives have ebbed in the community, its focus has shifted from cleaning out bad elements to bringing in beneficial ones' (Grossman and Leroux 2006).

If the economic and political forces driving gentrification hold sway, areas such as the Midsouth will be scrubbed clean 'of its working-class geography and history...its class and race contours rubbed smooth' (N. Smith 1996: 26–27). This is the significance of closing schools only to open them as new schools, dissociated from their former African-American working-class community identities. In their place are schools branded with qualities that appeal to middle-class tastes. For example, building on the success of private Montessori programs in Chicago's gentrified neighbourhoods, in April 2004, Arne Duncan, CEO of CPS, announced the opening of two new public school Montessori programs, noting, 'this is a great strategy to bring new families in [to CPS], middle class families' (Rossi 2004). Thus the contest over closing schools and opening new schools with new identities in their place is also about changing the meaning of the places where they are located and the struggle over identity and culture in the city (Keith and Pile 1993a, 1993b). This was implied by present and former students of Englewood High School, protesting their school's closing, who claimed, 'When you destroy a community's school, you destroy the community' (fieldnotes, 17 February 2005).

The dispersal and exclusion of low-income African Americans from a city in which their labour has become largely expendable and their presence is defined as 'dangerous', is facilitated by school policy. Before communities can be restructured, they have to be devalued, prepared for development, and reimagined as places of value; to do so it is necessary to construct a reality of 'easily discardable people and social life' (Wilson, Wouters and Cremmenos 2004: 1181). Race is central here. At a press conference announcing the closing of Englewood High School, CEO of Schools Duncan claimed that the school exhibited a 'culture of failure'—a racially coded marker of 'deficiency' in the Black community (Thorne 2005).

Although Ren 2010 is linked to the creation of new mixed-income communities, as noted above, the evidence suggests a small percentage of present low-income residents will be able to afford to live there. This is part of the neoliberal discourse of community renewal through private development that situates the problems of poverty and disinvestment in the supposed pathologies of low-income people of colour. As the middle class reinvents the city in its own image, 'the poor and working class are all too easily defined as "uncivil", on the wrong side of an heroic dividing line' (N. Smith 1996: 17).

A COUNTER-EXAMPLE

The 14 mothers and grandmothers who, in May 2002, set up tents on a vacant lot across from a cooking oil factory and began a hunger strike for a new high school in Chicago's Little Village community were joined by widespread community support. The demand for the school expressed a larger struggle for the welfare of the community. Latinos are the fastest growing demographic group in Chicago, comprising more than 26 percent of all residents and 34 percent of the city's children (Institute for Latino Studies). In the 2000 census, there were 750,000 Mexicans and 582,028 foreign-born Mexicans in the metro Chicago region in 2000 (Paral 2003). Many are immigrants who made the dangerous border crossing to escape the hardships produced by neoliberal economics in their own country and to seek a better life in the US. Here they face low-wage dangerous jobs, few social protections, racism, and discrimination (Davis 2000). Mexican immigrants have the lowest economic status in the Chicago metro-region (Paral 2003). Marginalized in low-wage jobs and concentrated in immigrant neighbourhoods, they nevertheless are present in the cultural life of the city and central to the city's economy. They are waging labour and community struggles for economic justice and against gentrification, and their growing numbers make them a looming political force in the US (Davis 2000).[7]

As defined by community activists, the struggle for the high school represents the quest for justice and respect for the community. Speaking about the city reneging on its promise to build a high school in Little Village, hunger striker Linda Sarate said, 'It was a slap in the face to the community, what they did.... It was like they were saying, "Your kids aren't worth it and my kids are"' (Russo 2003). A leader of the community organization that coordinated the planning for the new schools said, 'We went to visit the North Side magnet schools and they had all these facilities. We said that we want what they've got' (Russo 2003).

The school, which opened in the fall of 2005, symbolizes the strength and potential of the community. 'Just the fact that it is a new school brings hope', said one community leader (Russo 2003). At a cost of $63 million, it is the most expensive public school in the city's history and the most significant new structure in the community. All aspects of the building design, curriculum, and principal selection involved members of both communities. Architecturally, the school is constructed around 19 angles to commemorate the struggle that gave it birth. It includes a swimming pool, two gymnasiums, a day care centre, roof top auditorium, and a two-storey sundial that highlights the 19-day hunger strike. The complex houses four small high schools, one of which is the North Lawndale/Little Village High School of Social Justice. This school and another of the four is part of an emerging network of social justice high schools in the city striving to foster critical literacies and engage youth in progressive social action in their communities.

CONCLUSION: THE CONTEST OVER SCHOOLS IS A CONTEST FOR THE CITY

The transformation of Chicago is an instantiation of neoliberal economic and social processes that are reshaping cities globally. Stephen Gill (2000) argues that the present social conjuncture is characterized by global neoliberalism and resistance to it. This dialectic unfolds at multiple scales and in specific contexts, including education. Chicago demonstrates that education policy is not only shaped by global/local transformations, but that education is an arena of struggle over the constitution of urban space as a site of capital accumulation and cultural significance. Although gentrification is cast as a positive strategy for urban decay and the achievement of social stability, it is a means for the middle and upper middle classes to claim political, economic, and cultural control of the city (N. Smith 2002). Schools are central to this project in Chicago. Chicago illustrates that education policy can serve the 'class conquest of the city' (N. Smith 2002), but that it may also be a focal point of social struggles over control and meaning of urban space.

The debate about education unfolding in Chicago is part of a political struggle to define the city, to determine who will live where, who will benefit from the economic development taking place, and who will participate. Contention over Ren 2010 is on-going. In December 2004, because of community opposition, the Board was forced to withdraw the *Midsouth Plan*. However, in 2005, the Board moved to close several schools in the area, and opposition over school closing under Ren 2010 continues, not only in the Midsouth, but in Englewood and North Lawndale as well. A coalition of community organizations, teachers, unions, and school reform groups is pressing for a government moratorium on school closings under the plan.

Nor is the identity of the North Lawndale/Little Village High School and its relation to the communities it serves fixed. Political forces allied with the mayor are promoting a referendum to limit the school to the Little Village community, sowing divisions among African Americans and Latinos. This is significant because an African American-Latino alliance in the city would pose a political challenge to the current regime and its agenda. A school that embraces and unifies the youth and families of both communities is a potent symbol. Responding to the referendum, the principal of the High School for Social Justice, in an open letter to the parents, wrote: 'We believe in a school that can provide a safe, nurturing, and hopeful place for students from Little Village and North Lawndale. A place where these same students value, celebrate and affirm their cultural differences and similarities. A place where seeds of change will blossom and put an end to divisive political agendas' (Martinez n.d.). It remains to be seen whether the school's culture and curriculum will enact this vision. At the moment, it is a school in formation, developing in highly contested terrain.

How these issues will be resolved cannot be predicted. But they are instructive of the integral role of education in neoliberal urban development and resistance to it, including constructing new social relations of collectivity and inclusion and the right to justice in the city. A concrete analysis of 'actually existing neoliberalism' (Brenner and Theodore 2002)—its development in specific contexts, its contradictions, and the unfolding dialectic of resistance and counter-hegemonic projects—reveals that education is a field of power in which this dialectic plays out in struggles over the use and meaning of urban space.

ACKNOWLEDGMENTS

I wish to thank Nathan Haines for his contributions to the research and analysis of aspects of this chapter and the editors of this volume for their helpful comments. This paper could not have been written without the work and wisdom of the Kenwood Oakland Community Organization; Parents United for Responsible Education; LSC Chair, Adourthus McDowell; Teachers for Social Justice; members of the Chicago Teachers Union; and others fighting for equitable education in Chicago. All errors of fact and interpretation are mine.

NOTES

1. In a clear departure from a nationally decentralized system, NCLB mandated high stakes accountability measures, established sanctions for failing schools that include take-over by the states' education departments, and a prescribed menu of curriculum programs that must be adopted by failing schools.
2. Four small high schools—High School of Social Justice, Multicultural Arts School, Infinity Math Science Academy, World Languages—share the high school campus.
3. I attended and/or participated in monthly school board meetings. Four CPS public hearings, four rallies and pickets, two press conferences, 17 community organization and teacher meetings and forums, 11 coalition meetings, 20 planning meetings, and four congressional task force meetings. I had ongoing, sometimes daily conversations, with people at all levels of the school system and community involved in the issue.

4. To be eligible for replacement housing, residents must meet strict background criteria, work requirements, pass drug tests, and agree to follow behaviour rules that apply specifically for former CHA residents (Chicago Housing Authority 2003).
5. In my field notes from Board of Education meetings in 2004–2005, there is a consistent pattern of parents and teachers from African-American schools appealing for a working boiler, a new roof to replace one that is collapsing, an elevator to transport disabled students currently carried up stairs by school staff, windows that open, more classrooms, science labs, a cafeteria, books, etc.
6. This is reflected in two indicators: increase in housing prices and rate of house sales. These indicators were developed by Janet Smith, Urban Policy and Planning Program, University of Illinois, Chicago, http://www.uic.edu/cuppa/upp/people/faculty/smith.html. For the intersection of these indicators and Ren 2010 in the Midsouth see Lipman and Haines (2005).
7. On 10 March 2006, in protest at federal legislation that would criminalize undocumented immigrants and those who help them, 300,000 immigrants (by official estimates)—largely Mexican—staged the largest demonstration in Chicago in perhaps 100 years.

REFERENCES

Abu-Lughod, J. L.(1999) *New York, Chicago, Los Angeles: America's global cities*, Minneapolis: University of Minnesota Press.

——— (2005) 'Commentary: what is special about Chicago?', *City and Society*, 17.2: 289–303.

Almada, J. (20 February 2005) 'Englewood lots', *Chicago Tribune*, Sect. 16A: 1–2.

Betancur, J. J. and Gills, D.C. (2000) 'The restructuring of urban relations', in J. J. Betancur and D. C. Gills (eds.), *The collaborative city: opportunities and struggles for Blacks and Latinos in US cities*, New York: Garland.

Bluestein, B. (2005) 'A new business plan for Chicago. Executive Agenda, First Quarter', 2005, A.T. Kearney, accessed 12 April 2005, http://www.atkearney.com/main.taf?p=5,1,1,110,6

Brenner, N. and Theodore. N. (2002) 'Cities and the geographies of "actually existing neoliberalism"', *Antipode*, 34.3: 349–379.

Bryk, A. S., Sebring, P. B., Kerbow, D., Rollow, S. and Easton, J. Q. (1998) *Charting Chicago school reform: democratic localism as a lever for change*, Boulder, CO: Westview Press.

Catholic Charities of the Archdiocese of Chicago (1999) *The housing crisis in our neighbourhoods*, Chicago: Author.

Central area plan to help guide downtown growth (2002, July 2) City of Chicago, Department of Planning and Development, accessed 23 July 2002, http://www.ci.chi.il.us/PlanAndDev...PressReleases/centralareaplan.html.

Chicago Housing Authority (2003) 'CHA Plan for Transformation', accessed 24 September 2005, http://www.thecha.org/transformplan/plan_summary.html.

Civic Committee Press Release (24 June 2004) 'Chicago business leaders applaud Renaissance 2010—pledge financial and technical support', Civic Committee of the Commercial Club of Chicago, accessed 26 September 2005, http://www.lqe.org/News%20&%20Events/100NEWSchools-PressRelease.pdf

Commercial Club of Chicago. (2003) *Left behind, A Report of the Education Committee of the Civic Committee*, Chicago: Author.

Davis, M. (2000) *Magical urbanism: Latinos reinvent the US city*, London: Verso.

Dell'Angela, T. and Washburn, G. (26 June 2004,) 'Daly set to remake troubled schools', *Chicago Tribune*, Sec. 1.1, 22.

Geroulis, D. (10 August 2003,) 'Progress in Oakland may serve as model for city', Chicago Tribune Online Edition, accessed 5 May 2005, http://www.chicagotribune.com/classified /realestate/communities/chi-profile-69976.story

Gill, S. (2003) *Power and resistance in the new world order*, New York: Palgrave Macmillan.

Grossman, R. and Leroux, C. (2006) 'The unmaking of a ghetto', *Chicago Tribune Magazine*, January 29:11–16; 26–29.

Hackworth, J. and Smith. N. (2001) 'The changing state of gentrification', *Tijdschrift voor Ecionomische en Sociale Geografie*, 92.4: 464–477.

Handley, J. (2004, August 22) 'Redeveloping public housing', *Chicago Tribune*, sect. 16 C:1, 3A.

Harvey, D. (2001) *Spaces of capital: towards a critical geography*, New York: Routledge.

——— (2005) *A brief history of neoliberalism*, London: Oxford.

Haymes, S. N. (1995) *Race, culture and the city: a pedagogy for black urban struggle*, Albany: State University of New York Press.

Institute for Latino Studies (n.d.) 'The changing face of Chicagoland communities', Chicago Fact Finder, accessed 20 March 2006, http://www.nd.edu/~chifacts/chi_comm.html

Keith, M. and Pile, S. (1993a) 'The place of politics', in M. Keith and S. Pile (eds.), *Place and the politics of identity*, London: Routledge.

——— (1993b) 'The politics of place', in M. Keith and S. Pile (eds.), *Place and the politics of identity*, London: Routledge.

Life, F. (2005) 'Taxed to the limit', *The Chicago Reporter*, Nov/Dec., accessed 21 January 2006, http://www.chicagoreporter.com/.

Lipman, P. (2004) *High stakes education: inequality, globalization, and urban school reform*, New York: Routledge.

——— (2005) 'Metropolitan regions—New geographies of inequality in education: The Chicago Metroregion case', *Globalization, Societies, and Education*, 3.2:141–163.

Lipman, P. and Haines, N. (2005) 'From accountability to privatization to African American exclusion', Unpublished manuscript.

Lipsitz, G. (1998) *The possessive investment in whiteness*, Philadelphia: Temple University Press.

Martinez, R. (n.d.) Letter to Parents, accessed 20 March 2006, http://www.lvlhs.org/apps/news/show_news.jsp?REC_ID=12686&id=0&rn=1918.

Michaels Development Corporation, accessed 22 September 2005, http://www.michaelsdevelopmentcompany.com/portfRobertTaylor.html

Mendell, D. and Little, D. (2006, Feb 16) 'Change hits many inner suburbs', *Chicago Tribune*, Section 2:1.

Moberg, D. (in press) 'Economic restructuring: Chicago's precarious balance', in J. P. Koval, L. Bennett, M. Bennett, F. Demissie and R. Garner (eds.), *The new Chicago: a social and cultural analysis*, Philadelphia: Temple University Press.

Midsouth: Education Plan/ Renaissance 2010 Fact Sheet (n.d.). Kenwood Oakland Local School Council Alliance. Chicago, accessed 10 October 2004, http://www.teachersforjustice.org /pages/R2010factsheet.html.

Olivo, A. (6 May 2004,) 'Englewood rebirth plan brings hope and anxiety', *Chicago Tribune*, Section 1:1.

Orfield, G. (1990) 'Wasted talent, threatened future: metropolitan Chicago's human capital and Illinois public policy', in L. B. Joseph (ed.), *Creating jobs, creating workers: economic development and employment in metropolitan Chicago*, Chicago: University of Chicago Center for Urban Research and Policy Studies.

Paral, R. (2003) 'Chicago's immigrants break old patterns', *Migration Information Source*, Washington, D.C.: Migration Policy Institute, accessed 12 March 2004, http://www.migrationinformation.org/USfocus/

Perez, G. (2002) 'The other "real world": gentrification and the social construction of place in Chicago', *Urban Anthropology and Studies of Cultural Systems and World Economic Development*, 31.1: 37–69.

Popkin, S. J., Cunningham, M. K. and Woodley, W. (2003) Residents at risk, Urban Institute, accessed 25 March, 2004, http://www.urban.org/url.cfm?ID=310824.

Robertson, S. and Dale, R. (Draft 2003) 'Changing geographies of power in education: the politics of rescaling and its contradictions. EU funded GENIE–the Globalization and Europeanization Network in Education', accessed 11 January, 2004, http://www.genie-tn.net

Rossi, R. (2004, April 28) 'Chicago may get free Montessori', *Chicago Sun-Times*, Section News: 18.

Russo, A. (2003, June) 'Up Close: Little Village community organizing pays off', The Catalyst: Voice of Chicago school reform, Community Renewal Society, accessed 2 March, 2006, http://www.catalyst-chicago.org/arch/06-03/0603overview.htm

Sassen, S. (2002) 'Global cites and global city-regions: a comparison', in A. J. Scott (ed.), *Global city-regions: trends, theory, policy*, Oxford: Oxford University Press.

——— (2005) *Cities in a world economy* (2nd ed.), Thousand Oaks, CA: Pine Forge Press.

Schmid, J. (1998) 'City's loft, condo prices soaring', *Chicago Sun Times*, 4 September, N8-9.

Smith, M. P. (2001) *Transnational urbanism: locating globalization*, Oxford: Blackwell.

Smith, N. (1996) *The new urban frontier: gentrification and the revanchist city*, New York: Routledge.

——— (2002) 'New globalism, new urbanism: gentrification as global urban strategy', *Antipode*, 34.3: 427–450.

Soja, E.W. (1997) 'Six discourses on the postmetropolis', in S. Westwood and J. Williams (eds.), *Imagining cities: scripts, signs, memories*, London: Routledge.

——— (1999) 'In different spaces: the cultural turn in urban and regional political economy', *European Planning Studies*, 7.1: 65–75.

Thorne, K. (2005, February 1) 'Four Chicago area schools face uncertain future under Renaissance 2010', Medill News Service, Northwestern University.

Thrush Development Company, accessed 5 May, 2005, http://www.trushrealestate.com/html /jazzcommunity.htm

Venkatesh, S. A., Celimli, I., Miller, D., Murphy, A. and Turner, B. (February 2004) 'Chicago Public Housing transformation: a research report', New York: Center for Urban Research and Policy, Columbia University.

Wilson, D., Wouters, J. and Crammenos, D. (2004) 'Successful protect-community discourse: spatiality and politics in Chicago's Pilsen neighborhood', *Environment and Planning A*, 36:1173–1190.

10 On the right track

Railways and schools in late nineteenth century Sydney

Colin Symes

Policy shifts that valorise the market as a key predicate of educational provision have prompted research into schooling's cultural geography. There is now a better appreciation of the positional effects of schools, and the degree to which parents consciously swap schools, even localities to access more propitious educational environments. Failing that, their children undertake marathon journeys across a city to access favoured schools. Though such tactics are restricted to families in the know, those possessing the relevant information about schools, their consequences have been dramatic (Ball, Bowe, and Gewirtz 1995; Butler with Robson 2003; Hine and Mitchell 2003; see Taylor's chapter). They have led to a revival of selective high schooling, the abandonment of school zoning, and in Australia at least, a drift towards the private sector. In the 'social calculus' governing school selection, the idea of schooling children in their neighbourhoods, in the interests of producing more inclusive and democratic communities, which was a cornerstone principle of comprehensive education, has been abandoned. The cogency of the comprehensive ideal, which, even in the context of neoliberal policies still has many proponents, is one underpinned with the idea that placement of schools ought not to be taken lightly for it can promote social cohesion. In spite of this benefit, it was a policy that was rarely adhered to fully; there were always pockets of the population that chose, if they could, to avoid comprehensive education, either through accessing the selective remnants of the government sector or private schools (Connell 2003: 236; Connell et al. 1982; Sherington 2004: 175). This enabled them to preserve a way of life that countervailed the egalitarian imperative, whose privileges were dependent on the savvy stewardship of economic, social, and cultural capital. Self-assets such as education were integral parts of this stewardship and included attending schools that maintained this capital.

In this chapter, I argue that these strategies are not new; nor are they peculiar to Sydney (Barke 1991; Davison 1974; Dyos 1955; Fishman 1983; Pollins 1964).[1] Those of Sydney had their origins in the latter parts of the nineteenth century, during the first wave of the city's decentralization, when Sydney's middle classes moved to its burgeoning suburbia, seeded by the advent of the railways. The transformation wrought on the city's topography by the railways was powerful, and was underwritten with a supporting ideology that pressed home suburban exclusivity and with it, the separation of home from work along with other disruptions to social unity.[2] What was set in train was a class based form of segregation, which discouraged working families from 'going west' and limited their spatial horizons to the city's immediate precincts. Education in the form of prestige schools managed by opportunistic churches and entrepreneurs close to railways stations was a constituent part of this segregation and helped to ensure that the new suburbs were served with an education *system* matching the cultural and social aspirations of the middle classes (Sherington, Petersen and Bryce 1987: 64; Campbell and Sherington 2004: 8). Thus, the suburbanized middle classes were able to exploit to their advantage the 'power of place' (Hayden 1995) and isolate themselves from policies that might otherwise thwart their educational advancement (Marsden 1987: 73). That they continue to retain this advantage exemplifies the way the power of place once instantiated, is hard to dislodge. It also exemplifies the trenchancy of Massey's more temporal account of space, which emphasizes the interconnectivity of time and space; that considerations of time cannot be severed from considerations of space, and vice versa. The geography of power is as much a historical one as it is a spatial one, derived from the past as much as the present (Massey 2005). To her understanding of space, I add the factor of movement, which brings together time and space and recognizes the importance of mobility as a factor in maintaining and advancing social position. The spatial practices of the present have a complex genealogy, which I attempt to map. As I argue in this chapter the modernization of cities that occurred in the latter parts of the nineteenth century, and which was facilitated by the advent of the railways, left a lasting sociological legacy, which continues to haunt and disrupt the social efficiency of contemporary policies designed to offset the privileges conferred by geography.

ARCADIA INCARNATE

Australia's suburbanization was swift, exhibited most dramatically in Sydney. Its development as a city more or less spanned the nineteenth century, when urban planning was in the ascendancy and the establishment of transport networks transformed the urban morphology of the modern city. Initially, Sydney's residential development clung to its foreshore and was, for the most part, integrated with commercial and industrial activities. The almost tenfold growth of its population in the second half of the nineteenth century produced a congested, noisome, and unwholesome city, which gained a reputation as a cesspool of vulgarity and intemperance, disease and despair, disorder and debauchery (Mayne 1982: 79). Newspapers of the time drew attention to the city's high infant mortality rate, which was attributed to unsanitary dwellings, many deemed unfit for habitation by public health authorities (Mayne 1982: 104). It was soon to be followed by other acts to clean up the city, which by the 1920s had become a fully modern city: its streets were electrified, lined with skyscrapers and department stores, and there were many opportunities to embrace the latest fads in popular culture such as jazz and the movies (Matthews 2005).

With the recognition that foul air and crowded conditions were vectors for infection, public anxiety about the city grew.[3] Additional anxieties appertained to the amorality of the working classes: that licentiousness was rife, even among their children. Even though these claims were often hyperbolized ones, they bolstered middle class paranoia about the lower orders, helping governments to justify their surveillance and regulation (Mayne 1982). Fearing moral contagion from these dangerous populations, the middle classes undertook an exodus to the inchoate suburbs, whose development was underwritten by a long economic boom lasting from the 1830s to the Depression of the 1890s. Sydney's newly enriched gentry acquired large tracts of land on the city's fringes, on which they built large houses and mansions, before selling to speculators who subdivided their estates for further property development (Muir 1994: 17). Although the migration to more salubrious areas initially only involved the wealthy, it turned into an avalanche once Sydney's railway system was established. Arguably, it also set in motion an owner-occupier ideology, which has proved difficult to staunch, and resonates with neoliberalist morality, where individuals are encouraged to have proprietorship over self-goods such as education and health.

By the end of the 1880s, nearly four hundred thousand of the city's population were living in the suburbs and accessing the city via its trains, which had steadily increased in frequency and range throughout the 1860s and 1880s. Rail and the mobility it afforded were integral factors in Sydney's expansion, and helped set a pattern of development other Australian cities followed. Private investment in property was underwritten by government's development of infra-structure such as railways, which in turn produced further appreciation of property values, which encouraged owner-occupiers to invest in better housing stock, in remoter suburbs, made accessible by new railway lines, where land was cheaper and offered even greater returns on investment, *ad nauseam* (Daly 1982: 132). Investment in land and property emerged as a vehicle of bolstering social and economic position, as an alacritous means of acquiring wealth; hence, auctions of land in outer Sydney in the 1880s attracted wide public interest, which the Railways Department helped to excite, organizing special trains for the 'land hungry' (see Figure 10.1). Arguably, this pattern of urban expansion was repeated following World War II. It was spearheaded, as automobiles replaced trains as the main mode of mobility, by freeways rather than railway lines, and shows no signs of abating (see Kunstler 1994).

Fully-fledged suburbs with low population densities developed alongside the railway corridors. Proximity to a railway station, which tended to act

Figure 10.1 Land sale poster.

as a magnet of retail development, stimulated property values (Coupe and Coupe 1988: 98–99). The suburbs contrasted markedly with the congested inner city which was situated close to the railway lines, and served to remind commuters as to the reasons why they had moved to the semi-rural suburbs. These same suburbs also appealed to the growing enthusiasm for fresh air and nature.[4] The hopes and dreams of the aspiring middle classes were projected onto suburbia: it was the place where they installed themselves, where they began to feel at home (Harvey 2000: 158). However, Sydney's arcadia was not available to the working classes, who, at the time, were obliged to reside close to their places of work, in polluted, noisy suburbs, dominated by factories and industrial activity. Comparison with the present could not be starker: the former heartlands of the working classes, as has happened elsewhere in the world, have now been gentrified and are now desirable suburbs, occupied by the well-educated, professional classes.

THE LINES OF CLASS DIVISION

New South Wales' (NSW) rail system was planned with the transportation of freight in mind not passengers. But as Sydney expanded, its residents began to take advantage of its railways to access the city, which in turn spurred further development close to stations (Wortherspoon 1978; Fitzgerald 1987). Thus, the first corridor of suburbia followed the Western railway line, which opened in 1855, and passed through existing settlements, with stations located close to major roads (Hagarty 2001: 42). Running from Redfern terminus to Parramatta with intervening stations at

Newtown, Ashfield, Burwood, and Homebush, it catalyzed Sydney's western expansion. Within twelve months of opening, another line was added to the single line, and another two were added prior to the century's end. This was a vindication of the increasing demand for train services, which was also manifested in the construction of lines to the north and south of the city.

Prior to its subdivision, the inner west comprised market gardens, 'unfarmed' estates, and inns serving the Western highway. Aided and abetted by real estate agents, who encouraged the rich to flee the demonized city, the western suburbs soon became havens of pecuniary concentration, whose residents enjoyed the best of all possible worlds: the benefits of rural living and a link, courtesy of the new railway, to the city (Dunlop 1974: 36). Indeed, this link was one way developers 'sold' the suburbs. The residents of Ashfield—one of the first of the refugee suburbs—prided themselves on having, in sharp contrast to the profligate city, twelve churches and only five public houses (Fitzgerald 1987: 35). The vision of the suburb as a refuge from the debauched city took hold, such that by the end of the 1890s, 70 percent of Sydney's population was living outside the city, which furthered the demand for transport. On the city's western line, for example, several more stations were added, which in turn stimulated further residential development. The addition of Strathfield to the suburban network of stations was a direct result of the growing prosperity of the west, which by the late 1890s had become a 'municipality of mansions', centred on the recently subdivided Redmyre Estate, which emerged as a middle class enclave without peer in Sydney (Jones 1985: 46; Pluss 1985: 40).[5]

Thus, the advent of the railway facilitated Sydney's class division: the new suburbs created by the railway were, by and large, middle class, whereas the inner city remained predominantly working class.[6] The cost of railway tickets, which was high, along with the infrequency of train services, made the suburbs inconvenient locations for workers (Fitzgerald 1987: 52; Mayne 1981: 55, 60). Public transport, which included by the late 1880s an extensive network of tram services, was by and large scheduled and priced with the well-to-do in mind.[7] When the new timetable came into effect in 1880, there were eighteen services per day to Parramatta, fourteen to Homebush. There were no services on Sundays between the 'church hours'—a sign of the sabbatarian times (Isaacs 2005: 134). Even if they had capacity and/or the means to do so, the realities of commuting—expensive fares and unrealistic timetables—acted as brakes to the working classes migrating to the suburbs. Plus, they were actively discouraged from doing so, by being made to feel unwelcome, and out of place. Municipal councils backed by local ratepayers used their power to preserve the residential character of their suburbs and were often forced to abandon projects that were incommensurate with a suburb's demographic 'complexion'. Strathfield Council for example, opposed an extension of a tramway from Enfield to Druit Town, which would have allowed an influx of working classes into the suburb. The same council was forced to abandon a proposal to build an abattoir at Homebush in 1899. Local resident and lawyer Dr Sly argued it would ruin the area's residential character and healthy reputation, lead to lower property values and, worst of all, see an inward flux of undesirable residents (Fox and associates 1988). Notwithstanding Australia's democratic and egalitarian temper, the view was widespread that the new suburbs should be unashamedly middle class. One local resident observed that once the 'baker' was admitted into their midst, 'undesirable tenements' and 'miserable shanties' would be erected alongside 'our comfortable cottages or more ornate villa' along with the city's public house 'with its inherent vulgarity and vice, being thrown upon our wives' and daughters' morning promenade' (Green 1889: 152).

The social composition of Sydney's early suburbs then, reflected the realities of its public transport. Notwithstanding newspaper editorials calling for a reduction of fares and the introduction of worker friendly timetables, the railways remained the preserve of the middle classes.[8] It was these facts plus escalating land prices that placed suburbs such as Strathfield beyond the reach of the workers (Mayne 1982: 196–197). It was a pattern of settlement that continued as the railway extended into the city's northern suburbs such as Beecroft, whose residents in 1906 were lukewarm towards a proposal to schedule an early morning train for workers—not that they would have been able to afford the fare (Beecroft Cheltenham History Group 1995: 89). And this was after the long called for introduction of a workman's weekly ticket.

Although this ticket produced a steady increase of train patronage among workers, the increase was limited, by and large, to working class suburbs.[9] Yet it was also recognized, by no less a figure than the NSW Commissioner of Railways that the creation of new railway lines gave additional outlets 'for the overcrowded city' to grow (CfRT 1883: 11). It was a widely held view on the part of bureaucrats, then, that the development of the railways had assisted suburban growth; but the reality was that it was only the middle classes who were able to take advantage of this growth.[10] In the period from the 1860s through to the end of the nineteenth century the population of Sydney's inner west was an avowedly middle class one, and Anglican and Protestant at that. Following the depression of the 1890s, which caused a sharp downturn in property values, the middle class occupation of the inner west to some extent relented. The establishment of factories in neighbouring suburbs plus the development of more worker friendly transport services meant that living there became more plausible for working men and women. This was reflected in the subdivisions of larger properties in the district, which enabled smaller houses to be built in a price range attractive to workers. The gradual de-gentrification of the suburb prompted its own social backlash, which the extension of the railway to the North Shore helped facilitate.

In general terms, the significance of these observations suggests that by the latter parts of the nineteenth century the social effects of 'fluid modernity' (Bauman 2000; Kaufmann 2002) were well-understood by Sydney's middle classes who had begun to exploit the spatial annihilating effects (Kern 1983: 214; Thrift 1996: 264) of transport and to deploy mobility as a factor of their social calculus, using the removal from the city to more communally homogeneous suburbs as an opportunity to buttress their economic and social capital.[11] In effect, they recognized that the train's capacity to reduce travel time, could be mobilized to secure more control over urban space; that transport provided more freedom of movement across a range of social fields, including education (Urry 2003: 171). This served to normalize movement from place to place on a daily basis as an inescapable element of contemporary life. It also produced particular envelopes of 'time-space', a circadian cycle of movement, in and around the city, the 9–5 of workplace, the 9–3 of schools and so on (Crang 2001).

None of this occurred overnight, but was worked out over several generations and involved spatial manoeuvres designed to reinforce social privilege. Part of knowing one's place was having the foresight to undertake those manoeuvres and knowing when to make them; more to the point, it involved needing to overcome the temptation to remain stationary. None of this was achieved in a socially neutered or benign way but, as Latour (1997) has argued, required the enlistment of new forces of labour to work for the travelling publics and guarantee fluidity of movement around the city and eventually across, and between, countries.

GOVERNMENT ACTS: THE ADVENT OF STATE SCHOOLING

It was to address the fragmentary and sectarian nature of the colony's schooling that the New South Wales government, beginning in the 1860s, undertook a series of policy reforms. These had the effect of bringing denominational schools under the control of the Council of Education in return for which they received a government subsidy; they also allowed communities to petition the council to provide a school in their districts. This was provided they could ensure at least 25 enrolments, and could contribute at least a third of the costs towards its building, which favoured suburbs with the requisite pecuniary assets. The most forthright expression of educational modernization in New South Wales was the proclamation of the 1880 Public Instruction Act, which instigated a system of free, compulsory, and secular instruction for students. The fact that the Act made elementary schooling compulsory entailed the undertaking of a massive school building programme across the state, and which the advent of the railways, paradoxically, both exacerbated and ameliorated. In 1890, the Minister of Public Instruction recognized that the 'rapid extension of the railways' along with the 'increased populations extending along these railways' had placed 'great demands for expenditure for new schools' (RMPI 1891: 84). On the other hand, the existence of an extensive rail system assisted in making the new schools more accessible for those children, living in country areas that were close to regional centres. It was in order to take advantage of this connection and to foster an increase in state school enrolments that free transport was provided for children attending state schools. This was an entitlement not extended to those children attending non-government schools, which further exacerbated the sectarian tensions already evident in the state's education system (Barcan 1965: 184–185). Notwithstanding these tensions, the reforms produced their intended results, and there was a 22 percent shift of pupils from the private- to the state-sector (Chambers 1986: 77). The advent of free schooling that the 1880 Act instigated represented a defeat of sorts, albeit a temporary one, for free enterprise schooling.

Although the architects of the 1880 Act were mainly concerned with establishing universal primary education, they also prescribed a limited programme of secondary education across the state. This occurred in two ways: through separate high schools established in Sydney and the towns of Maitland, Goulburn, and Bathurst; and in a series of nominated Superior Public Schools in remote areas, which offered secondary education in primary schools. This far-sighted decision arose not so much from public demand as the recognition that entrance to Sydney University from 1866 onwards depended upon passing an entrance examination, for which government schools, with the exception of Fort Street Model School, were ill equipped to prepare students.

By and large, then, the commitment to secondary schooling was a pusillanimous one, reflecting the view that properly speaking secondary education should be left to the private sphere. The fact that the two Sydney government high schools initially occupied a building—boys on one floor, girls on the other—that had been originally set aside for a library and had to be modified (Dunlop 1965: 65), reflected this. Even with these modifications, the two schools were soon overcrowded and the minister began lobbying for a more 'spacious site' (RMPI 1889: 20).[12]

The bush foray into secondary education fared no better, and failed to attract sufficient students. Of the original six rural high schools, only Maitland survived more than a few years. Those at Goulburn and Bathurst, which were located in rented premises, faced stiff competition from local 'colleges and ladies' schools' and soon closed. Contributing to their premature demise were at least three factors: first, were the fees charged, which were uncompetitive in the educational market of the times; second, was the fact that private schools provided 'certain accomplishments' absent from the state curriculum; third, was the poor standard of their facilities (RMPI 1885: 150). Another factor was that they failed to have much appeal for middle class families: an 1889 report noted that while 'all grades of society are represented' in government high schools they draw the bulk of their pupils from families 'who have to work for their living' and 'deny themselves much in order to give their children the higher education that these schools afford' (RMPI 1889: 161). In other words, the middle classes had begun to recognize the social and cultural importance of education and saw that what the government provided—even though its academic performance was impressive—only partially provided their children with the social and cultural accomplishments they needed. State schools were seen as a second class, residual, in today's terms, attended only when means precluded anything better. Indeed one mealy-mouthed critic of the time, seemed to think that a general meanness appertained to education, that would drive parents, at the merest hint of a straitening of their financial circumstances, to 'maim the education' of their children (Singleton Spalding 1889: 315).

The state high schools also faced internal competition from the superior public schools, which charged lower fees yet achieved comparable academic outcomes (Dunlop 1965: 64; Kyle 1986: 112). Even though they were conceived with remote locations in mind, by the end of 1890s there were several Superior Public Schools located in Sydney's inner west, which proved popular, so much so, that by the early 1900s, the local inspector contemplated placing an embargo on enrolments (RMPI 1901: 74). The government also faced opposition from the Catholic Church, which thought that the state's foray into schooling was misguided, and sought to undo its further expansion. Then there were the early advocates of an educational market (see also Reid 2005). They argued that, in subsidizing schools, the government had embarked on a course of protectionism that risked undoing the law of supply and demand, which had operated in the private sphere. Subsidies would 'crush out of existence' schools doing an adequate job (Bean and Sly 1886).[13]

But this did not happen. The fee debacle combined with the failure of high schools to attract pupils, suggested that the government's commitment to secondary schooling was in large measure a rhetorical one, that is, until the second decade of the twentieth century, when Peter Board, the Department's new secretary, abandoned fees, and undertook a programme of establishing high schools throughout the state. Arguably, this was too late to undo the middle class loyalty to private schools, which by this time had had a generation to prove their social and educational efficacy, and to capture enough support to guarantee their survival into the future.

PRIVATE EDUCATION AND SYDNEY'S INNER WEST

Suburbs are protean spaces, always subject to social construction and reconstruction, as their communities seek, through civic action, to influence their development and consolidate their character.[14] Entrepreneurs also participate in this process of reconstruction, as they attempt to take advantage of the market stimulated by new housing. The suburbs of the inner west were no exception: stylish shops, sporting clubs, music and drama societies proliferated, reinforcing the roots of privilege and respectability. They provided the contexts in which the freemasonry of the local bourgeoisie was established and organized, and its political power marshalled.

In the case of schooling, this power operated on two fronts: public and private. As younger families moved into the railway suburbs, they petitioned, with the support of the local inspector, the government to build a public school, as happened, beginning in 1879, at Croydon. In April 1881, the inspector's observation that the people of the district 'belong to the professional and artisan classes, and are in comfortable circumstances' seems to have sealed the case. But even before this happened, opportunistic real estate agents, acting on behalf of local property owners, had contacted the Council of Education with offers of land, that, given its proximity to a station, it was said would make ideal sites for government schools.[15]

As Strathfield's reputation as an enclave of prosperity grew, school entrepreneurs, in the main drawn from the ranks of the clergy and well-educated women, felt that they could provide an education more commensurate with their class needs than the state (Barcan 1965: 159).[16] They attempted to take advantage of the education middle class families were seeking, and who recognized, the importance of full education in maintaining the social station of their children, and who wished to avoid them 'socializing' with their state school counterparts.[17] Not that they would have had much opportunity to do so. From the 1880s through to the 1920s, there was a vacuum of secondary education in Sydney's inner west, which the government, perhaps bruised by its bush experiences, seemed reluctant to fill (Sherington, Petersen and Brice 1987: 5).

What secondary education there was outside the inner city then, was, in the main, privately run or operated by various church denominations and took the form of small schools with enrolments of twenty or so students (Dunlop n.d.: 57). The schools were not avowedly religious, of which many parents, reflecting the secular times, approved, but were as organizational entities vulnerable to fluctuations of the educational market and the whims of the teachers; hence, they gained a reputation for impermanency. By the end of the nineteenth century, though, the Anglicans, Catholics, Presbyterians, and the Wesleyans had entered the picture and had opened a number of schools that were devoid of this problem. The smaller schools went into steady decline. By the 1920s, by which time they required government registration and depended for their survival on students eligible under the Bursary Endowment Scheme of 1904, their numbers had dropped rapidly.

Before moving to Stanmore in the 1880s, Newington College—the first high school established in Sydney by the Wesleyans—was sited close to Parramatta, at a relatively remote location, Silverwater, accessible only by 'paddle steamer' or after a long walk from Lidcombe (then Haslem's Creek) station (Swain 1999). Declining enrolments and the fact that the school had outgrown its Silverwater location prompted relocation to an area where the school could expand; where it could also take advantage of the day-school market. Its proximity to a station meant that it was accessible to boys from across Sydney. Newington was the first educational foray into New South Wales by the Wesleyans. The relocation to Stanmore—then a fast developing suburb—had been prompted by the fact that the Methodists and Presbyterians, following on from similar initiatives in Melbourne (Sherington, Petersen, and Bryce 1987: 47), had already established two girls' schools, which had facilities for boarders, in the area: Methodist Ladies' College (MLC), at Burwood, and Presbyterian Ladies' College (PLC) at Croydon. When in 1881 Sydney University admitted girls as students, the demand for a rigorous academic education for them grew quite rapidly and it was this demand that MLC and PLC were intent on satisfying, thereby entering into direct competition with Sydney's Girls' High, which had already acquired an enviable academic reputation. It is of note that in MLC's case, location was a consideration: the specifications presented to the sub-committee examining a site for Sydney's MLC included that it be close to a railway station and no further out than Burwood (Coupe and Coupe 1986: 6). This proximity was deemed desirable not just for daygirls, but also the boarders, which the school hoped to attract.

As a counterforce to its state competitor in the city, the non-Conformist schools made their appeal through advertising in national and regional newspapers. Just prior to MLC opening, an advertisement for the school in the *Town and Country Journal*, as well as promoting the school's academic virtues in regard to making 'provision for those who wish to prepare for University honours', also claimed a number of positional benefits for the school. One related to the school being 'accessible by train from Sydney'—a regular theme in school advertisements of the time (see Figure 10.2).

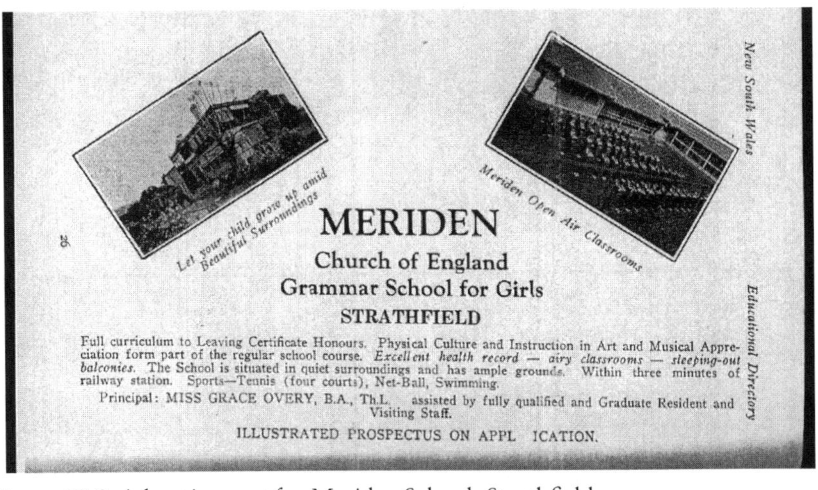

Figure 10.2 Advertisement for Meriden School, Strathfield.

Indeed, according to the school's prospectus for 1900, the school's hours were timed in accordance with local train timetable.[18]

In the case of Santa Sabina, a Strathfield Catholic girls' school belonging to the Dominican order, which opened in 1894, the selection of a suitable site had proved difficult. The school's planners had teased with several sites, including Dr Sly's Eton College, then on the market. Far better was its existing site, five minutes walking distance from Strathfield station, considered ideal because it was half way between Maitland and Moss Vale (also on the railway line), where the Dominicans also had interests (Emilsen 1994: 18). And when the Sydney Church of England Grammar School (SCEGS) contemplated, before dropping the idea, purchasing what subsequently became Strathfield's Meriden School, one mark in favour of its purchase was the school's closeness to the western and northern lines (Cooke 1997: 24).

Proximity to a railway station then was a determining factor in school placement (Seddon 1985). Ill-placed schools often experienced enrolment declines and would relocate to redress the problem, only to experience the same problem later on in their histories if the demographics of their new locations changed for the worse (McCulloch 1992; Marsden 1987). A school's attraction was seemingly enhanced if it was located near to a station in a desirable location, which was why it frequently featured in school advertising (Figure 10. 2). The explanation for this was that the ability to travel across the city was a socially restricted one. A school which attracted its students from across the city, parents inferred, must have a good reputation. By contrast, those drawing students from their immediate zones were less attractive, particularly if they were located in working class suburbs (Marsden 1987: 73).

Other advantages related to the school's topography. Schools in semi-rural settings ('Let your child grow up amid Beautiful surroundings'), particularly those commanding a view, it was felt provided a more 'natural' learning environment (Seddon 1985: 191). Of equal importance, though, was the fact that the school was sufficiently distant from 'the keen air of the sea' and was situated in an 'eminently healthy neighbourhood'.[19] Thus, many schools exploited in their advertising prevailing fears about epidemics in the city, and for which the inner west's healthy atmospherics provided some protection ('Excellent health record—airy classrooms'). Such appeals carried prudential weight and encouraged parents who had fallen into line with the ventilation ethic, then a dominant part of educational thinking (see Wilmot and Saul 1998), to shift their children from the city to the suburbs.

The same advertising also alluded to another educational trait of the times: that of schools taking over existing schools, in this case, Miss Lester's, and using the credentials of the existing school to reinforce the new school. In the cut-throat educational market of late nineteenth century Sydney, private schools, which no longer enjoyed financial protection from government, rarely survived more than one generation, and often went bankrupt or were taken over by their larger, more financially adroit rivals.

Another trait of the time was for schools to occupy, when they became vacant,[20] the palatial mansions that dotted the district, and then adding extra educational buildings as and when required. This was the case with Presbyterian Ladies' College (PLC), which in 1891 took over the former home of the shopping magnate, Anthony Hordern III, on a site adjacent to the railway line, and which continues to dominate the PLC campus, an ever-present structural reminder of the school's privileged origins. It is as though the architectural 'traditions' of the home continue to underwrite the philosophy of the school, which is to create a class apart from the educational run of the mill, based on the conflation of the ancient and modern. This was different from state schools, which were built from scratch, from the ground upwards, and demolished all that was around them. They were purpose built with the educational endeavour in mind and deployed the latest architectural insights into the optimization of space, to maximize the efficacy of modern pedagogy. Such matters as the volume and size of the classroom and the amount of space to be allotted to the each child were all, in line with emergent sciences of administering populations, matters for prescription.

By the 1890s, educational investment in the inner-west had begun to diminish. It had fulfilled its purpose, which was to provide the suburbs with the educational machinery required to maintain their 'classical' purity. With the exception of Santa Sabina, no new private schools were opened. This was partly because of the onset of economic depression, which had been precipitated by plummeting prices for Australia's exports, along with profligate government investment in public works programmes such as railways. Suburban growth had peaked, hoisted itself on its own petard, and urban expansion all but ceased for the next twenty years (Fry 1978: 40–41).

NEW EDUCATIONAL HEIGHTS ON THE NORTH SHORE

One of the remarkable features of Australian cities is that they have continued to grow into the undeveloped hinterlands around them (Fry 1978: 32). The middle classes were, by and large, the first to colonize these hinterlands, and once having done so moved on to somewhere else to colonize. The NSW government underwrote this activity by providing the development of appropriate infra-structure: transport, sewage, schools, and water supply. In 1886, for example, the suburban railway system was extended through to the North Shore, first from Strathfield to Hornsby, and then, in 1890 back to Sydney via an exclusively suburban line to St Leonards with a later extension to Milson's Point, which had the effect of creating new land for suburban development, in an area now encompassing the suburbs from Roseville through to Wahroonga. The residents of the North Shore had long pressed for their own railway line and felt aggrieved that the NSW government had not listened to their pleas. The government felt there was not sufficient public benefit to be gained from building such a line, and money spent on developing the north would be better expended on the booming suburbs in the south and the west. Speculators already owned large tracts of land alongside the proposed railway line and stood to gain financially when the land was resumed, and this did not help their cause (Jones 1988: 151). When the government finally relented, large tracts of native bush and forest were clear felled and transformed into Sydney's second suburban utopia. The private school 'cartel' had, anticipating the shift northward of Sydney's middle classes, already established itself there, thus pre-empting any state initiatives in the area. For example, Barker College moved from a relatively remote location in the Blue Mountains to Hornsby; following the success of its Croydon venture, a branch of Presbyterian Ladies' College was opened at Pymble in 1916. It was following the precedent of Methodist Ladies' College in this regard which in 1901 had purchased a large girls school in Gordon for a branch of its Burwood campus. These schools all sought to take advantage of the fact that their campuses were on, or close, to the North Shore railway line. This was also true of Knox Grammar and Abbotsleigh, which were both located at Wahroonga and opened in 1924, the latter moving from its Parramatta site to take advantage of an airier location on the North Shore. Indeed, all these schools were upwardly mobile in more senses than one; for as their prospectuses proclaimed they made much of the fact that they were not only within minutes or a 'few yards' in the case of Knox, of their respective railway stations but also situated on a site 'about six hundred feet above sea level', which meant they were 'favoured with a fine climate' and were subject to the health giving properties of altitude.

CONCLUSION

In this chapter, I have argued that the development of Sydney's inner west followed the course of its railway system, which prompted an exodus of the middle classes from the inner city. I have shown that various interests took advantage of suburban development: developers, churches, retailers, owners of sporting clubs, and so on, and which operated in conjunction with one another to establish and then preserve the class ambience of the area. The production of space is indubitably linked to producing communities of compatibility, such that individuals come to know where their place is, and are comfortable living there. The result in this instance was the establishment of a ghetto of privilege that was partially dependent on the burgeoning railway system that provided the travelling infrastructure of Sydney's first suburbs. It was also dependent on a schooling system commensurate with the long-term cultural and social interests of the community. Thus from 1880s onwards the inner west became Sydney's educational heartland, and although many of the schools that were established during this period soon closed, the ones that survived are now the leading bastions of corporate schooling in NSW. The location of these schools near to railway stations gave them a distinct positional advantage in attracting students from across Sydney, which they continue to enjoy. Indeed, in terms of the gentrification of its inner city, these schools have experienced an enrolment renaissance as middle class families have removed their children from inner-city government schools, which are now predominantly working class and ethnic ones (see Gulson's chapter).[21] Of more significance though, is the fact that the development of the schools in the inner-west 'normalized' the private school as the natural organ of secondary education for the middle classes and undermined any possibility that state education could ever be the primary site of educational provision for them. This had long term consequences, mostly negative, for the provision of schooling in New South Wales, which the recent upsurge of neoliberalist approaches to schooling have further exacerbated. The argument of this chapter then, supports the notion that the grammar of space is driven by past, present, and future imperatives, and that the roots of social disadvantage have their tentacles entrenched in space. This reinforces the propositions of Massey: that time and space must be thought of together, not as a closed system, but as factors that work one with the other, that are forever being reworked and worked out, that are mutually necessary in the never ending chain of social life (2005: 14). In Sydney at least, in the nineteenth century the middle classes were better able to anticipate and to harness to their advantage the processes of urban modernization that the advent of the railway underpinned. They took up the challenge of the railway and arranged their spatial behaviours accordingly via the increased fluidity provided by trains. More to the point, its challenge had a long-term impact on the social patterns of schooling in the city—something policy makers ignored in the provision and take-up of schooling. Always, the educational geography of the present is inherited from the past, and while the negative impact of that inheritance persists there is little prospect of generating a more egalitarian and inclusive system of education. That is why space needs to be taken seriously.

NOTES

1. The suburban imperative though is not characteristic of European cities, such as Paris, or Latin American cities (Fishman 1987: 114), where the urban elites have for the most part remained in its 'Haussmanized' heart. Nonetheless, part of Haussman's justification for the redevelopment of Paris was predicated on stopping the intertwining of the classes (Sennett 1986: 134).
2. Many commentators of the time, John Dewey among them, were troubled by the city's mechanization, which was seen as fostering 'contactless' interactions between citizens (Schwarzlose 1984: 88). Indeed, devices such as the telephone made suburban living more viable because they enabled residents to

remain in contact with their work without having to be in the city—another instance of action at a distance.

3. This anxiety was realized when epidemics of typhoid and plague swept across the city in 1880 and 1900; nor were the suburbs spared.
4. There were two sources for this enthusiasm: first, was English Romanticism, which worshipped nature (see Harris and Larkham 1999: 10); second, was the Evangelical Movement, which valued the spiritual revitalization flowing from contact with nature (Fishman 1987: 56).
5. Strathfield's residents included businessmen, politicians, judges, and archbishops (see *Some notes on the Municipality of Strathfield*).
6. Sydney guidebooks of the time described the class composition of suburbs. Samuel's *How to know Sydney* (1897: 72) described Alexandria, Annandale, and Balmain as 'working men's suburbs' and Ashfield as a 'residential suburb'. Interestingly, The Tourist's Bureau's *Excursion Map of Sydney and surroundings together with a concise guide to all the principal places of interest, public buildings, etc., etc.*, first published in 1887, offers a thumbnail sketch of Burwood, which is pictured as a 'large and fashionable railway suburb'—as if proximity to the railway was a mark of distinction. At the same time, some new suburbs, such as Daceyville, Sydney's first purpose built garden suburb, were assigned to the working class.
7. In 1891 a first class single fare from Ashfield cost 6 pence, from Strathfield 8 pence, which the Commissioner for Railways at the time claimed compared favourably with equivalent fares in England (CfRT1894: 10). Yet, the same difficulties were evident in London (Pollins 1964).
8. See *The Sydney Morning Herald*'s editorial for 13 February 1884, which suggests that the cost of second class fares was excessive—the first especially so. Indeed, the system's first historian commented that a return trip to Parramatta 'must have been regarded somewhat in the light of a luxury to be enjoyed by the rich only' (Gilder 1905: 25). There had been a third class, which was abolished in 1864. An article in *The Sydney Echo* notes the uselessness of the northern railway for working men because the first train from Ryde left at 8.12 a.m. and that they would need to walk 2 miles to Homebush to get to the city in time for 'their daily avocation' ('Suburbs of Sydney, No. XXII—Strathfield and Homebush', 18 September 1890)
9. In 1883 for example, the bulk of such tickets were sold at stations serving working class suburbs such as Macdonald Town and Newtown. Twenty-five percent fewer were sold at Summer Hill, Ashfield, and Croydon (CfRT 1883: 135).
10. Speaking to a public meeting, the Mayor of Strathfield noted that the proposed line north of the municipality would 'open up the district' and 'Concord would advance as rapidly in settlement and population as Burwood, Homebush and other railway suburbs had done' ('The Mortlake Tramway: proposed line north of Strathfield', *The Australian Courier* 11 April 1891.
11. The idea that space could be annihilated was first adumbrated in the middle of the nineteenth century (Report of the Commissioners 1839).
12. The boys were moved in the 1880s to Ultimo, and the girls to Moore Park in 1921.
13. One of the authors of the pamphlet, J. D. Sly, eventually established Eton College in Homebush, which was finally crushed out of existence by market forces: the school went bankrupt!
14. The residents of Strathfield petitioned their council to plant trees along its main streets, thus helping to strengthen the suburb's reputation as a garden suburb.
15. Typical was Property Auctioneers Hardie and Gowan, who were acting on behalf of Mr Hardie. Another property owner in the district, Anthony

Hordern, also approached the Council but the local inspector thought that schools at Ashfield and Burwood could accommodate the Croydon children. There was some suspicion that the property owners were seeking prices that were in excess of market value and that it would be in the interests of the government to resume rather than purchase the land (see State Records New South Wales Croydon Public School 1876–1897, Bundle A)

16. Some fifteen private schools and colleges are listed as located at Ashfield, and Petersham. See *Sand's Directory for Sydney and NSW* for 1890.
17. As the educational classifieds in *The Sydney Morning Herald* show, these schools were located across the city, wherever there were pockets of middle class families, e.g. Darlinghurst, Woollahra. Some schools would move their locations to more propitious sites, e.g. Scots College was originally located at Rockdale before moving in 1893 to the more middle class Bellevue Hill.
18. I am grateful to MLC's archivist, Mylee Angelo, for this observation.
19. See *Town and Country Journal* 2 January 1888. The school is advertised under its then name, Wesleyan Ladies' College.
20. This was especially true after 1906, when the minimum wage case ensured that the employment of domestic servants—an indispensable requirement for the upkeep of these mansions—led to their demise as family residences and their conversion into schools and nursing homes (Fox and associates 1988).
21. 2005 figures compiled by the New South Department of Education and Training (DET), suggest that there has been a slight reversal of this trend, and city state schools are experiencing enrollment resurgences.

REFERENCES

Ball, S., Bowe, R. and Gewirtz, S. (1995) 'Circuits of schooling: sociological explanation of parental choice of school in social class contexts', *Sociological Review*, 43.1: 52–78.

Barcan, A. (1965) *A short history of education in New South Wales*, Sydney: Martindale Press.

Barke, M. (1991) 'The middle-class journey to work in Newcastle upon Tyne', *Journal of Transport History*, 12.2: 107–134.

Bauman, Z. (2000) *Liquid modernity*, Cambridge: Polity.

Bean, E. and Sly, J. D. (1886) *High schools versus scholarships: an enquiry into the merits of the two system*, Bathurst: Glyndwr Whalan.

Beecroft Cheltenham History Group (1995) *Beecroft and Cheltenham: the shaping of a Sydney community to 1914*, Beecroft, NSW: Beecroft Cheltenham History Group.

Butler, T. with Robson, G. (2003) *London calling: the middle classes and the remaking of inner London*, Oxford: Berg.

Campbell, C. and Sherington, G. (2004) 'The public comprehensive school in New South Wales', *Change: transformations in education*, 7.1: 1–16.

Chambers, C. (1986) *Lessons for ladies: a social history of girls' education in Australasia 1870-1900*, Marrickville, NSW: Hale and Ironmonger.

Commissioner for Railways and Tramways (CfRT) (1883) Report for the year 1882, Sydney: Government Printer.

—— (1894) *Report for the year 1894–3*, Sydney: Government Printer.

Connell, R. W. (2003) 'Working-class families and the new secondary education', *Australian Journal of Education*, 47.3: 235–250.

Connell, R. W., Ashenden, D. L., Kessler, S. and Dowsett, G. W. (1982) *Making the difference: schools, families and social divisions*, Sydney: George Allen and Unwin.

Cooke, A. (1997) *Vision of Parnassus: Meriden's first 100 years*, Strathfield, NSW: Meriden School.

Coupe, S. and Coupe, R. (1986) *The walk in the light: MLC school Burwood*, Burwood, NSW: MLC School Burwood.

Crang, M. (2001) 'Rhythms of the city: temporalised space and motion', in J. May and N. Thrift (eds.), *Timespace: geographies of temporality*, London: Routledge.

Daly, M. T. (1982) *Sydney boom, Sydney bust: the city and its property market 1850–1981*, Sydney: George Allen and Unwin.

Davis, A. (1930–1931) *The Australian educational directory and gazette and compendium of information on public examinations*, Sydney: Minerva.

Davison, G. (1974) 'Public utilities and the expansion of Melbourne in the 1880s', in C. B. Schedvin and J. W. McCarty (eds.), *Urbanization in Australia*, Sydney: Sydney University Press.

Dunlop, E. W. (1965) 'The public high schools of New South Wales, 1883–1912', *Journal of the Royal Australian Historical Society*, 51.1: 60–86.

—— (1974) *Harvest of the years: the story of Burwood 1794–1974*, Burwood: Municipality of Burwood.

—— (n.d.) *Between two highways: the story of early Croydon*, Surry Hills, NSW: Wentworth Press.

Dyos, H. J. (1955) 'Railways and housing in Victorian London: II. Rustic townsmen', *The Journal of Transport History*, 11.2: 90–99.

Emilsen, S. (1994) *Dancing St Dom's plot: a history of Santa Sabina and Santa Monte, Strathfield*, Strathfield, NSW: Santa Sabina College.

Fishman, R. (1987) *Bourgeois utopias: the rise and fall of suburbia*, New York: Basic Books.

Fitzgerald, S. (1987) Rising damp: Sydney 1870–90, Melbourne: Oxford University Press.

Fox, M. and associates (1988) *Strathfield heritage study*, prepared on behalf of Strathfield Municipal and the Department of Environment and Planning.

Fry, E. C. (1978) 'The growth of Sydney', in J. W. McCarty and C. B. Schedvin (eds.), *Australian capital cities: historical essays*, Sydney: Sydney University Press.

Gilder, G. (1905) *Fifty years of making: a historical essay on the New South Wales railways from their inception*, Sydney: Building Printing Works.

Green, J. (1889) 'Model suburbs—II: Harcourt, Burwood', *The Centennial Magazine*, 2.2: 150–154.

Hagarty, D. (2001) 'The railways at Summer Hill', in C. Pratten (ed.), *Summer Hill*, Ashfield, NSW: Ashfield and District Historical Society.

Harris, M. A. (1917) *Where to live: ABC guide to Sydney and suburbs*, Sydney: Marchant.

Harris, R. and Larkham, P. (1999) 'Suburban foundation, form and function', in R. Harris and P. Larkham (eds.), *Changing suburbs: foundation, form and function*, London: E and F.N. Spon.

Harvey, D. (2000) *Spaces of hope*, Berkeley, CA: University of California Press.

Hayden, D. (1995) *The power of place: urban landscapes and public history*, Cambridge, MA: MIT Press.

Hine, J. and Mitchell, F. (2003) *Transport disadvantage and social exclusion: exclusionary mechanisms in transport in urban Scotland*, Aldershot: Ashgate.

Isaacs, V. (2005) 'New South Wales railways passengers services 1880–1905', *Australian Railway History*, 56.810: 132–157.

Jones, M. (1985) *Oasis in the west: Strathfield's last hundred years*, North Sydney: Allen and Unwin.
—— (1988) *North Sydney, 1788–1988*, Sydney: Allen and Unwin.
Kaufmann, V. (2002) *Rethinking mobility: contemporary sociology*, Aldershot: Ashgate.
Kern, S. (1983) *The nature of time and space 1880–1918*, Cambridge, MA: Harvard University Press.
Kunstler, J. H. (1994) *The geography of nowhere: the rise and decline of America's man-made landscape*, New York: Touchstone.
Kyle, N. (1986) *Her natural destiny: the education of women in New South Wales*, Kensington, NSW: New South Wales University Press.
Latour, B. (1997) 'Trains of thought: Piaget, formalism, and the fifth dimension', *Common knowledge*, 6: 170–191.
Marsden, W. E. (1987) *Unequal educational provision in England and Wales: the nineteenth century*, London: Woburn Press.
Massey, D. (2005) For space, London: Sage.
Matthews, J. J. (2005) *Dance hall and picture palace: Sydney's romance with modernity*, Sydney: Currency Press.
Mayne, A. J. C. (1981) 'Commuter travel and class mobility in Sydney, 1858–1888', *Australian Economic History Review*, 21. 1: 53–65.
—— (1982) F*ever, squalor and vice: sanitation and social policy in Victorian Sydney*, St Luica, Qld.: University of Queensland Press.
McCulloch, G. (1992) 'Constructing the community: secondary schools and their neighbourhoods in 20th century Auckland', *Australian Journal of Education*, 36.2: 143–156.
Muir, L. (1994) 'Shady acres: developers and the design of Sydney's public transport system 1873–1891', Ph.D. diss, University of Sydney.
Pluss, M. (1985) 'The evolution of Strathfield', in I. Burnley and J. Forest (eds.), *Living in cities: urbanism and society in metropolitan Australia*, Sydney: Allen and Unwin.
Pollins, H. (1964) 'Transport lines and social divisions', in R. Glass et al. (eds.), *London: Aspects of change*, London: MacGibbon and Kee.
Reid, A. (2005) 'The regulated education market has a past', *Discourse: studies in the cultural politics of education*, 26.1: 79–94.
Report of the Commisioners (1839) 'Report of the Commissioners appointed to consider and recommend a general system of railways for Ireland', *The Quarterly Review*, 68: 1–60.
RMPI (1885) *Report of the Minister of Public Instruction for the year 1884*, Sydney: Government Printer.
—— (1889) *Report of the Minister of Public Instruction for the year 1888*, Sydney: Government Printer.
—— (1891) *Report of the Minister of Public Instruction for the year 1890*, Sydney: Government Printer.
—— (1901) *Report of the Minister of Public Instruction for the year 1900*, Sydney: Government Printer.
Schwarzlose, R. A. (1984) 'Technology and the individual: the impact of innovations on communication', in C. L. Covert and J. D. Stevens (eds.), *Mass media between the wars: perceptions of cultural tension, 1918–1941*, Syracuse, NY: Syracuse University Press.
Seddon, G. (1985) 'Some notes on the geography of education', in I. Palmer (ed.), *Melbourne Studies in Education*, Melbourne: Melbourne University Press.
Sennett, R. (1986) *The fall of public man*, London: Faber and Faber.

Sherington, G. (2004) 'Public commitment and private choice in Australian secondary education', in R. Aldrich (ed.), *Public or private education? Lessons from history*, London: Woburn Press.

Sherington, G., Petersen, R. C. and Bryce, I. (1987) *Learning to lead: a history of girls' and boys' corporate secondary schools in Australia*, North Sydney: Allen and Unwin.

Singleton Spalding, J. (1889) 'Our high schools', *The Centennial Magazine*, 2.4: 313–317.

Swain, P. L. (1999) *Newington across the years: a history of Newington College 1863–1998*, Stanmore, NSW: Newington College.

Thrift, N. (1996) *Spatial formations*, London: Sage.

Urry, J. (2003) 'Social networks, travel and talk', *British Journal of Sociology*, 54.2: 155–175.

Wilmot, F. and Saul, P. (1998) *A breath of fresh air: Birmingham's open-air schools, 1911–1970*, Chichester: Phillmore.

Wortherspoon, G. (1978) 'The 'Sydney Interest' and the rail 1860–1900', in M. Kelly (ed.), *Nineteenth-century Sydney: essays in urban history*, Sydney: Sydney University Press.

11 Student mobility and the spatial production of cosmopolitan identities

Michael Singh, Fazal Rizvi, and Mona Shrestha

It is difficult to understand the internationalization of higher education in recent decades without linking it to the geo-political changes borne of the spatial expansion of global capitalism. Universities continue to be affected by 'neoliberal' state-driven policies directed towards accretive privatization, socializing possibilities for technological proliferation and the intensification of work-place regulation through international competition (Morrow and Torres 2000). The state-driven valorization of the regulatory force of quasi-markets appears to have become an arbiter of the educational values of universities. National governments have used international conventions and agreements to subordinate public education policies to the ideological project of 'neoliberal globalism'. Once the education of the public was intended to deepen and extend the socio-economic security of sovereign citizens through forming and informing their character and virtue, as much as the constitution of democratic nation-states. But now, the state appears to have become a vehicle for producing a different kind of subject. Increasingly, students are expected to take advantage of the opportunities offered by the global expansion of higher education, while graduates are expected to have the capacity to see, imagine, and experience labour market possibilities beyond the nation-state.

The redefinition of universities as providing a marketable commodity reflects and expresses the changed political relationship between the state, labour markets, and citizenship. This spatial reorientation has favoured the extension and hierarchical differentiation of universities and their governance structures. The drawing of higher education into these markets through internationalization of its services has also facilitated the continuous re-structuring (and de-structuring) of students' perceptions, conceptions and lived experiences of universities. Following Berland (1992: 50), it might be argued that the effects of these changes are 'to expand the political and spatial reach and legitimation of corporate interests and initiatives in conjunction with the growing impoverishment and disenfranchisement of much of the population'. The mobility of students internationally in this, the latest phase(s) in the long-term transition of globalization, is linked, in a range of complicated ways, to the socio-economic contradictions inherent in expanding the spatial reach and trans-national connections of universities.

This chapter examines the experiences of international students in their journey back and forth between different places. Particular attention is given to how their perceptions, conceptions, and experiences of these spaces contribute to the production of their 'cosmopolitan' sense of belonging. It explores how their identities are elaborated by their negotiation of spatial awareness, ideas, and practices. As Morley and Robins (1995) argue, because places are no longer internally homogeneous, bounded areas, identities are increasingly constructed out of material and symbolic resources which reach beyond local boundaries. Through trans-national mobility, international students imagine, conceive, and experience being insiders and outsiders in both their places of origin and destinations. In doing so they contribute to the production of a distinctive cosmopolitan space mediated and mitigated by the cultural turbulence of globalizing forces, connections, and desires. The opportunities provided by trans-national mobility form and inform shifts in their growing sense of cosmopolitanism.

The concept of 'space' is central to our analysis, as we explore its potential in analyzing how international students' identity shifts over time, from their conception of life abroad, through their perceptions of life overseas, to their lived experiences at home and away. We draw on Henri Lefebvre's (1991) critical work on the production of space to argue that the spaces perceived, conceived, and experienced by international students are strategically negotiated by them, as they seek to overcome challenges to take advantage of the possibilities presented by the globalizing economy and culture. The dynamics associated with these processes elongate their richly textured cosmopolitan identities.

Taking space as lived and imagined practices, as much as symbolic representations, the analysis in this chapter is based on evidence generated from interviews with international students from China who studied in Australia. Four specific themes provide the focus for this chapter. First, the sense of otherness generated by language and cultural barriers provide them with the threshold for negotiating a space in which to secure a sense of belonging. Second, space is shown to be integral to the shift in balance between the challenges of day-to-day survival and building a sense of independence and opportunities. Third, there is the dilemma of generating a sense of belonging (or not) in the quest for global cultural spaces, as students seek to create new spaces that are both vehicles for promoting global uniformity as much as they are sites of division and diversity. And finally, the new spaces produced by the students are not neutral and contain markers of class formations and privilege, with their mobility becoming a major element in their sense of cosmopolitanism. But first we consider a little of what has been written about the concept of space.

THEORIZING SPACE

In recent decades an absolute objective understanding of space as a system of organization or geometry has been largely abandoned (Massey 1994; Soja 2000). Space is much more than a structural grid within which objects are located and events occur. In analytical terms, space is no longer treated as a container of static, though movable objects and dynamic flows of behaviour. The dimensions and contents of space are no longer assumed to be natural and given. It is less likely that space is measured solely in terms of pre-existing physical laws. Critical geographers such as Pile and Thrift (1995) contend that the search for general spatial laws is futile. Such laws explain little about the interrelationship between people and place, and do little to help identify the logic in patterns of human settlement and relationships. Indeed, as Crang and Thrift (2000: 2) argue, absolute and essentialized conceptions of space have paradoxically reduced the world to 'spaceless abstraction'. Such conceptions pay insufficient attention to the 'spatial problematic' elaborated upon below.

In contrast, a relational view of space seeks to provide a relative understanding of how space is constituted and given meaning through human endeavour. Space is no longer given a neutral or passive geometry; it is continuously produced through socio-spatial relations. The relationship between space, spatial forms, and behaviour is no longer assumed to follow a set of natural spatial laws. Instead, space is conceived as a product of cultural, social, political, and economic interactions, imaginings, desires, and relations. In this way, space is not merely an objective structure but also a social experience. As Massey (1994: 254) notes, space is constituted through social relations and material practices. This does not mean that space is entirely a social construction.

In a very helpful analysis partially informed by the work of Lefebvre, Soja (1985) makes a useful distinction between space and spatiality, suggesting that while not all space is socially produced all spatiality is. His analysis is thus focused both on the symbolic construction of space at the level of social imaginary as well as its more concrete articulation in the landscape. In this way, a university, for example, can be minimally interpreted as a spatial allocation that has physical form. Perhaps, more importantly, a university is a complex phenomenon given meaning through myths, language, and rituals that speak to its spatial form. Universities may be defined in terms of their socially defined spatial arrangements, such that the terms 'international students' or 'off-shore campus' give expression to their changing purposes, range of relationships and processes.

Writing in the early 1970s, Henri Lefebvre (1991) cautioned against the cult of words, textual reductionism, and the fetishization of discourse, to provide a persuasive account of how space is produced through three interrelated processes of spatialization. First, he has identified a set of spatial practices through which it is possible to identify flows and movements in the realm of everyday routines. 'Where' university students pursue their academic endeavours is influenced by their journeys through and in space. Second, he has spoken of a set of 'representations of space' in images, books, media, and so on. These serve to represent and make sense of space. Such representations have the capacity to reproduce space, working ideologically to legitimate or contest particular spatial practices. Maps of university campuses can thus be recognized as social productions that purport to offer an objective view of their increasingly multiple localities. Finally, Lefebvre has identified 'spaces of representation'. This is the space that is lived, felt, and experienced as students negotiate their way through the contingencies of everyday life, as they look back to their past experiences and express their aspirations and make future plans. These spaces are imbued with ideological and political content. They involve dealing with broader structures, including various contrasting representations of space and spatial practices.

Although Lefebvre's (1991) complex relational view of space and spatiality has a thin presence in education research (Robertson, Bonal and Dale 2002; Danaher, Danaher and Moriarty 2003), it underscores the importance of human agency. It points to the connection between macroeconomic and geopolitical transformations, and the networks of social action that students create, move in, and act upon in their daily lives. At least it suggests a need to account for the ways in which human agents interpret, engage with, and negotiate these generalized processes, at the level of their own constructions, intentionality, everyday practices, social relationships, and collective action. This does not mean that investigating the 'agency' of students is a transparent social process. On the contrary, it involves the mess and confusion of actually living in today's very different world. There are difficulties associated with attempts to sort out the contingencies inherent in moving from subjectivity to deliberative human agency. Despite this, it is nonetheless important not to reify global structures as pre-given things existing outside thought. Despite the recurring predictable critiques of the logic of neoliberal rhetoric, neither can fully explain what international students are up to. How they understand and represent the spaces that they occupy requires the study of their efforts to re-negotiate these spaces, as they develop new social networks and a sense of trans-national imaginary.

Our discussion in this chapter is located within this 'spatial problematic'. It draws upon data collected as part of a major project designed to address a range of issues concerning global mobility, especially the contemporary consumerist production of mobile identities in and through international education. The project examined how international students from China perceived, conceived, and experienced their education in and through Australian universities (Rizvi 2005). It involved the study of how their experiences of mobility, transition, and travel, as well as how their formal and informal learning stretched their emerging cosmopolitan identity (Singh and Sproats 2005). This project investigated their cultures on the one hand and the emergent global culture(s) on the other. More specifically, the project sought to explore the various ways in which international students *used* their experiences in Australia to think about their cultural affiliations and possibilities (Singh 2005). This study explored the challenges they faced in (re)inserting themselves into their own communities and in using their education in building professional careers. In this way, we looked at both their pre-departure experiences in China, their insertion into Australia, and their experiences back in China after graduation. In this chapter we discuss these experiences specifically in spatial terms. We assume that space is not fixed but is in a 'state of becoming', and can only be known in and through both time and mobility.

In framing our research questions in this way, we rejected the view of both space and cultural identity as something fixed, stable, and unchanging. Cultural identities are not located within a pre-determined objective space that reflects common historical experiences and shared codes that provide continuous frames of reference for making meaning. Instead, we worked from an alternative perspective. While we recognize that identities come from somewhere and have histories, we nonetheless maintain that they are under constant transformation. In this respect, Stuart Hall's (1996: 225) observation that 'far from being eternally fixed in some essentialized past, they are subject to continuous "play" of history, culture and power' was particularly poignant to us. This perspective highlights active processes of meaning-making—of identity formation. The dynamics of mobility, transculturalisms, and 'diasporization' are notions that underscore these continuous processes of spatial production of cosmopolitan identities.

With this emphasis on agency, we preferred the term 'uses of international education' rather than the notion of 'outcomes'. The latter is employed widely to explore the graduate destinations and career trajectories of international students. The focus on the 'uses of international education' is designed to mark a philosophical and methodological departure from the functionalist idea of 'outcomes'. We adopted the term 'uses of international education' based loosely on Richard Hoggart's (1958) notion of 'the uses of literacy' and Cameron McCarthy's (1998) idea of 'the uses of culture'. It has enabled the exploration of the processes of cultural *production* through which international students struggled to make sense of their experiences in Australia; the ways in which they assessed their histories in China and, how they imagined and anticipated their future spatial trajectory; and the ways in which they felt they were positioned by and actively sought to position themselves within the dominating corporate narratives of globalization and international education. In this way, our research interest related to each of Lefebvre's (1991) three complementary levels of theorizing 'space' in terms of spatial practices, representations of space, and spaces of representation. However, we have done this without giving analytical priority to one over any of the others because they have a complex, unstable, and interpenetrating relationship.

The research involved interviews with 80 students from China, both before they completed their programs in Australia and also upon their return. In what follows, our focus is on the spaces conceived, perceived, and lived by international students from China as they were positioned, and as they positioned themselves as mobile, global consumers in the worldwide university market that has a dual character as a market for higher education commodities and as a market for trans-national labour immigration. We seek to show how their day-to-day practices of lodging and travelling, processes of spatial production provide a way of accounting for shifts in their identity formation. Our argument is that these international students came to Australia with already formed spatial imaginaries of 'home' and 'abroad', both of which were dislodged during their life in Australia and upon their return, and that together these perceptions, conceptions, and experiences of space were fundamental to their attempts to enlarge their cosmopolitan identities. However, in tracing the shifts in their sense of identity relative to the changes in their sense of space, we suggest that this cosmopolitanism is of a particular kind. It is informed by their social obligations which give them a sense of belonging to a certain home-town or nation-state; the spatial reorientation of corporatized universities; as well as the trans-national opportunities for migration and work within the emerging structures of global capitalism.

MAKING SENSE OF AUSTRALIA

Our interviews show how, in seeking to make sense of the social and cultural dynamics they encounter in Australia, both within the university and beyond, the international students negotiate their sense of space. Their sense of space changes as they explore the constraints and possibilities their lived experiences abroad produce. The case of Wen Ming shows how his experiences of Australia and Australian higher education in the late 1990s involved a range of considerations relevant to the ways he represented and made sense of space. When we interviewed him in 2002, Wen Ming worked in the public affairs department of a private company in China. He had graduated from a Shanghai university with a Bachelor of Engineering in 1989 before working for a government organization for two years, followed by the State Economics Trade Commission for six years. He received a World Bank scholarship to study for a degree in economics in 1997 at an Australian university. When he returned to China he continued working with the State Commission until internal governmental reforms led him to join a private company.

When Wen Ming went to Australia for his studies his initial encounters were melancholic. While at that time it was still too expensive for him to make frequent telephone calls back home, the first three months he nonetheless had to make several calls a week to overcome his anxieties, borne out of his feelings of homesickness. As the student grapples with the reality of taking root in another space, it is the memory of home that drives this obsessive nostalgia. Standing in opposition to a wandering existence, the home is a place where dreams and rationality can dwell in virtual 'ontological dignity' (Lefebvre 1991: 121). This emotional, 'maternal if not uterine' bond is such that everywhere people's home is 'a special, still sacred, quasi-religious and in fact almost absolute space' (Lefebvre 1991: 121).

To overcome such feelings of homesickness, he realised early that he needed to be pro-active in creating a new space for himself which might connect him to students with similar experiences. Ming was determined to make new friends, transforming his new 'home' into a space of lived experience, governed by the practicalities of his studies and the character of the people with whom he sought interactions. In retrospect, he feels his adverse situation provided him with a stimulus to his learning, especially his personal development. Also, it created a space constituted by a social network of friends with whom he could share the meaning he attached to his experiences in Australia. In this sense, space was not his own individual construction but a social formation. While the space of his university consisted of work units (schools, faculties, centres, offices), these friendship networks were also part of the space for his education. The contours and boundaries of the corporatized university provided one networking of space, among many interpenetrating spaces that were articulated with, but not necessarily governed by its solid but fragile neoliberal imperatives.

Upon his return to China, Ming had to once again forge new spatial practices in knotted and stratified networks. He encountered many practical difficulties in relating his new knowledge and newfound sense of professional identity to the system and culture in his place of origin. Nevertheless, he saw huge opportunities in China for those like himself who had studied abroad. Both the multinational companies operating in China and the Chinese government needed the expertise of graduates of foreign universities, but the space occupied by an earlier era had not disappeared. Lefebvre (1991: 403) argues that 'no space disappears completely, or is utterly abolished in the course of the process of social development.... "Something" always survives or endures'. People such as Ming were needed to grow the nation's burgeoning economy and enable China to become globally integrated. Even so, Ming now found he needed to work with colleagues who 'think-alike'. He found spaces, the networks and links, which would enable the specific exchanges and uses he had imagined before leaving China to study overseas, and that his education there had produced. He felt a need to work with other people who had been educated abroad. He formed a group in his company to fulfil this desire while maintaining the network of contacts and friends from his student days. These local/global relations served both social and business purposes. Once again these represented his changing sense of the space in which he lived and worked.

The international students we interviewed regarded global mobility as a threat as much as an avenue for opening up new horizons. Such spatial movement is something that could be productively or unproductively consumed, given how easily its benefits might be worn out by rubbing against things that have endured from the past, at home and abroad. In Australia, the students reported feeling excluded from public constructions of identity or found themselves up against regressive forms of localism, even racism. This bedrock historical space known as 'White Australia' is the product of the 'bonds of consanguinity, soil and language' (Lefebvre 1991: 48) and has survived despite urgings for its demise which have simultaneously prolonged its life. Huang Bo, now an academic in Nanjing, expressed his experiences of being an alienated body in an unwelcoming space, aware of being gazed upon:

> I was a foreigner…it's just strange feelings. I can say that some people in Australia are not very friendly to the people from other countries. I had one experience. One day I was waiting for the bus—there was a middle aged Australian who asked me where I'm from and what I was doing there. I told him and he said, 'Well, we don't like you very much because you come to Australia and take our money'. That did not happen very often. I just had one experience but that made me very unhappy because I thought I'm a hard working person. One of my roommates told me he had the same experience.

For these international students, public transport provided but one venue for the performance of White Australian angst. The students experienced an unwelcoming space that they had to negotiate. They had to make use of this lived space, with its hardships because it was part of their everyday activities for getting a higher education. The tensions that they experienced in this lived space were integral to their maturing cosmopolitan identity. Their failure to be able to deal with these conflicts would leave their capacities in these areas underdeveloped. Lefebvre (1991: 362) makes the point that 'lived space bears the stamp of conflict between inevitable, if long and difficult maturation process, and a failure to mature that leaves particular original resources and reserves untouched'. The students had to produce their own sense of space in which they could be comfortable.

Another student, Xiaowen Wang, shared Wen Ming's self-awareness of an imperfect foreign ear in an accentuated space. International students are grounded in a language-specific space, they cannot absent themselves from this localization of space or concepts, nor do they want to be excluded from these. Lefebvre (1991: 132) observes that 'every language is located in space. Every discourse says something about a space (places or sets of places); and every discourse is emitted from a space'. Xiaowen mapped herself into the linguistically sanctioned regulations governing the blending of World Englishes:

> The first week when I arrived it was at that time I did think my English was OK. By the end of the first few weeks, I said, 'Wow! I don't understand'. I went to the shops and to mail letters and all this. But because, you know, it was the accents of the Australians. I totally could not understand for the first few weeks, but after that I started to be OK.

It was not always possible for these international students to give the impression that they were blending in. In various ways in dissimilar locations they were scrutinized by others and pressed to ask who they themselves were. In terms of the relations of inclusion and exclusion, these international students did not relate to this space as they might when contemplating a picture or enjoying a spectacle. They knew that they were enveloped in multi-layered spaces which they had to actively engage: '[t]hey know that they *have* a space and that they *are* in this space... they act and situate themselves in space as active participants' (Lefebvre 1991: 294, original emphasis). Wen Ming noted that their identities were revealed at those times when they became aware of the regulations that alienated them from locals:

> Australian students get a discount for public transport, but for all international students there are no such concessions. We have to buy full fares. That's not good. When I was first here, I felt much better because we were treated the same but now I think they treat us differently. This is not good. Although they make some money from the travel fees they charge, maybe they lose more money in other ways.

What Wen Ming's observation reveals is his consciousness of higher education becoming an international marketplace. The spatial arrangement of students locally and globally occurs in a seemingly linguistically indifferent and culturally impersonal space. This market indifference led these international students to realize that they needed to re-create a new space for themselves. It was this repressed emotion that created their sense of existential unease which could be either a nostalgic desire for all that was imagined to be home or an entry point of negotiation (Morley and Robins 1995: 25). With this sense of wistful comparison and the feeling of otherness, a threshold was established to generate a shared space among those who were in a similar situation. Huang Bo explained:

> I lived with other students in my house—there were two from Sri Lanka, one from Japan and one from Thailand. I think we had the same feelings. We were foreigners. Well, I think it's good. All exchanges of ideas and experiences are good for...open discussion.

A bridge among these students from different countries was established. However, in creating this space it also left them with the constant reminder that together they were 'others'.

These international students had to both interpret and engage with the spaces created for them by the university and their cultural encounters with the broader Australian community. At the same time they had to negotiate a new space in which they could achieve a sense of belonging with others, a feeling of inter-connectedness within and outside their imagined communities. In this sense, the production of space is a dynamic process incorporating practices, values, and beliefs shaping students' capacities to make and re-make their identities. However, the space into which they moved was not merely a passive backdrop. Wen Ming, Huang Bo, and Xiaowen Wang struggled 'to achieve the ability to make their appearance blend in, in different ways in different places, under the scrutiny of the gaze and graze of the other, under the self-scrutiny of the mirror' (Pile and Thrift 1995: 46). They masked those elements of their identities that threatened to expose them as unblended. These differences were revealed in moments when they were not sure what to say or how to behave. They were never sure as to what local students might think of them. Suddenly, they would feel self-conscious or alienated. While these international students mapped their selves into socially-sanctioned spaces, they did not do so perfectly.

THE SPATIAL CHALLENGES OF SURVIVAL AND INDEPENDENCE

Their imperfect mapping of their bodies into locally grounded spatial practices was integral to the opportunities that the international students we interviewed had to experience the 'deterritorialization of culture' (Tomlinson 1999). This enabled them to produce cosmopolitan identities, borne out of the dual challenges of survival and independence. One of the important reasons for international students to venture out to new and unknown territories is their sense of independence. With their strong aspiration to be self-reliant they are able to cope with their daily struggles to blend in. Wen Ming provided a striking example of this:

> For the first three months when I called my wife I was really homesick. I thought a lot of things were so different from everything in China. After that it got better, I started to have more friends. Because of the Christian society I met a lot of people. As I became familiar with everything I really liked the city, the university, everything. Very positive. Probably I changed. I changed a lot through the programs, through the studies, having so many friends and also attending social activities—totally changed me, yeah.

For Wen Ming resilience did not come easily. Social space is not susceptible to ready decoding as it is over-inscribed in multifarious and overlapping ways. This was more a state of becoming rather than being. In a similar fashion, Xiang Li, an information technologist, explained her struggle thus:

> I worked part-time as a teachers' aide. All the way through I was the only one. One or two students were working part-time in my class. When my Dean knew I was working in part-time, he just could not believe I had sustained my studies. He said there is no way you can work part-time as well as commit to your study. I said, but I have to. And I did. I think I did pretty well.

To succeed both academically and professionally these students had to decipher the invariably confusing messages they read into the several codes expressed in this space. This they had to do against its prohibitions and dissimulations for doing so. This space was not 'produced in order to be read and grasped, but rather in order to be lived...' (Lefebvre 1991: 143). Li Ling, a telecommunications officer, depicted a similar determination:

> I remember when I first arrived I experienced some excitement. I went through some stages of being very undecided and worried because when I was in China, everything was planned by the society or by Dad. When I went to Australia, all of a sudden, I had to make decisions. I had to work—I had to do something, right? Thanks to my previous education, I was able to speak English in a way that I was able to mix with the people there. Of course, the first thing I decided to do was to make some money and to work. In that way, I started to make friends... I worked very hard. I worked at weekends; I worked during the holiday time basically to support myself.

It was the determination to survive that helped carry Li Ling and Xiang Li through the rough periods. It is this achievement of survival and becoming independent that was noted throughout our interviews. The students viewed the 'deterritorialization of culture' as both a challenge as well as an opportunity to become independent 'cosmopolitans'. Lefebvre (1991: 289) observes that there are 'beneficiaries of space, just as there are those excluded from it, those deprived of space'. These students, the beneficiaries of new configurations of local/global space, would now agree with Bauman's (1998: 2) observation that 'being local in a globalized world is a sign of social deprivation and degradation'. Despite the alienation they experienced in Australia, their views echoed the proposition that 'localities are losing their meaning-generating and meaning-negotiating capacity and are increasingly dependent on sense-giving and interpreting actions which they do not control' (Bauman 1998: 2).

Everyday the students engaged in future-oriented, coping strategies. However, it cannot be said that they moved deliberately through their studies or experiential learnings in ways that necessarily or decisively realized their best intentions. The students' 'comportment' (Pile and Thrift 1995: 28) indicated their openness and adaptability. Their orientation was to take all activities as providing learning opportunities. They were surprised when things went wrong; but they made deliberate efforts to deal with things that went wrong. Even so, the students' own intended actions did not decisively guarantee the shape of their resulting learning and work/life trajectory. Competition and other displacing forces affected these outcomes in ways that cannot be attributed to their own strategic desires.

In this sense the students had generative or creative capacities. For them space was not static, fixed, or passive. Their university was a place of desire and fear as it invoked:

> a different kind of space between the person who looks and the object that looks back: there is a position, distance and an orientation to the look, which specifies a particular space of meaning and power: this space is neither isolated nor abstract; this space both contains and refuses an infinite number of invocations of meaning and power; this space is constitutive of the visual practice, it is staging and integrating lines of power and meaning between the look and the look-back. (Pile and Thrift 1995: 46)

In the spatialization of their experiences, the centre of gravity shifted from the question 'Who are you?' The more important questions that emerged were, 'From what intersecting points in planetary space are you making your observations?' 'Which knowledge mesh-works are you connected into?'

Immobile and mobile students do not occupy a similar local/international placement. Therefore they are unable to observe the world and the practices of globalization from the same standpoint. This means that not all sightings by mobile students are likely to be of equal value or equally valued by immobile students, and vice versa. This generates pedagogical struggles over whether there must, should, or could be a certain privileged standpoint from which the 'best' education, the best knowledge can be gained. The internationalization of education, with its chaotic and bewildering complexity, has not made it easy to decide what 'best' now means. The idea of forming one universal world image to be shared by all students asserts a hierarchy of (superior) worldly knowledges, languages, and imaginings.

NEGOTIATING GLOBAL CULTURAL SPACES

Upon their return to China, the international students we interviewed had to once again negotiate the space they occupied. Their transformations in

Australia meant that they could not easily settle back into the comfortable local or national space they had once inhabited. These young people expressed little nostalgia for either the imaginings or reality of the 'actually existing socialist modes of production' of decades past. Their spatial practices had changed, sometimes to the extent that their behaviour was not really appropriate to the local culture. Wen Ming noted:

> Because I worked with these colleagues for many years, I know these people, so after I came back I didn't feel it was very difficult to communicate with them. But I felt that my mind needed to switch back so quickly—the ways of dealing with people...in China, in the government especially, people have to protect themselves, so this is the difference. So when I was in Australia, it was very open. I tried to be open but I felt, 'Oh, I cannot say that...I decided okay there are some things I do not say unless I think very clearly.

These students did not want to flee (post)modern global cities rather they wanted to engage the difficulties of life in a globalizing world. It is reasonable to say that they were developing their capacity to conceive of, deal with the contradictions and to treat space on a worldwide or global scale, and to do likewise through their direct experience of the fragmentary sub-divisions of space evident in a multiplicity of disjointed university procedures or processes (Lefebvre 1991: 355). Zhao had a similar story of forming and struggling with plural identities:

> Another advantage of my studies abroad was I could see something from different angles. Previously, even here in China, we used to think of something in a Chinese way. Now sometimes I'm thinking in the Western way... It's very complicated. If you're employed at a Chinese factory directly—it's too complicated for Chinese workers—they have to modify something like that. Even the idea. You have to modify some of the ideas you have.

Both Wen Ming and Zhao felt that their Australian experiences had made them culturally more flexible. In a fundamental sense they were now bilingual, able to be more reflexive about how they had to negotiate spaces differently. They were not ones for preserving representations of communist spaces or maintaining its arrangements unchanged, but keen to be agents of hyper-industrialization and hyper-urbanization.

They had also recognized that spaces are not static, that forms of transnationality were feeding into their continuous transformation. Jun Fan, a city planner, explained her new found mediated global cultural space:

> I can make Japanese sushi, Indian curry and an Australian barbeque—a lot of things—I can make pizza for my son.... So I think the life

> in Australia, within the two years, was pretty exciting and was very worthwhile. When I came back to China, sometimes I have party... stay together and talking...I have the same feeling. The two years, the studying and the living have somehow changed my life.

Her Australian experiences had helped to develop an indomitable reflexive capacity. She faced her world not 'as a set of significations to be interpreted [but rather] as a concrete set of problems that [she had] have to solve practically' (Pile and Thrift 1995: 15).

Li Ling displayed a similar reflexive disposition. Apparently, her formal education did not give her the strategies to negotiate, reinterpret, and remake the materiality of her circumstance. However, her informal experiences in Australia did so. She engaged in the process of decipherment, moving from obscurity to enlightenment as her reflections pierced the surface to penetrate the depths of these spatial relationships (Lefebvre 1991: 182). She was now able to negotiate both local and global spaces simultaneously. This is perhaps the manner in which she expresses her cosmopolitanism. As she remarked:

> I have been away now from Australia for quite a few years, I have continued constant contact with those people in Australia, not only from those professional networks as well as friends I made... I think the reason I keep the relationships is basically because nowadays when I am now in China, it's like every time friends come from Australia, I have the feeling that there is someone from my home town coming... Because to me now there are no boundaries between countries any more and especially for people like in our case.

Li Ling's sentiments cut across the dualism between home and abroad. Just as the national borders are becoming increasingly more porous (at least for some), so too is people's sense of national belonging. They have now trans-national spaces which they negotiate for themselves. As global economic and cultural production and consumption become increasingly integrated, so too does the desire for cosmopolitanism by international students—perhaps even more so upon their return rather than while they were in Australia. Against the familiarity of multiple cultural spaces, and with the confidence of their own skills of cultural negotiation, they now seek new orientations to their world and new forms of cosmopolitan bridging. This does not necessarily mean cohesion and unity but a global space that is expressed in its plurality mediated by trans-national friendships and professions. For the students we interviewed, their international education had increased the scope and potential they have for elaborating new forms of bridging: 'new sense of community; new attachments and allegiances; new identities and subjectivities' (Morley and Robins 1995: 5).

THE SPATIAL PRODUCTION OF COSMOPOLITAN IDENTITIES

Through this chapter's limited engagement with Lefebvre's (1991) treatise on the production of space, the discussion above indicates how during their stay in Australia and upon their return 'home', these international students continually negotiated their spatial practices, representational spaces, and spaces of representation. Through their international education they engaged in spatial practices that enabled them to live in cosmopolitan spaces produced not only by their mobility but also by their experiences of biculturalism. These international students represented themselves in these spaces neither as immigrants to Australia, nor as tourists. Their representational space embodied complex symbolism unavailable to tourists, because they were not in Australia for a short period, interested only in an outsider's gaze. Nor did their representational space embody the complexities of immigrants, even if many had an intention to settle in Australia permanently. Both the concepts of tourism and immigration represent border-maintaining distinctions exerted around and by the nation-state. For these students Australia offered spaces of representation related to the production of trans-national knowledge workers (Han 2004).

As Elden (2004: 86–105) suggests, remaining exclusively with Lefebvre's (1991) ideas on spatial production (and 'Western' conceptual tools more generally) has drawbacks given the importance of trans-national mobility of different peoples to critical conceptualizations of space. As cultural theorists have pointed out (Clifford 1997; hooks 1992), in the current phase(s) of globalization, these distinctions are becoming overwhelmed by cultural and technological innovations. These changes have swept across frontiers, contributing to the declining capacity of the nation-states to maintain these dimensions of their borders.

The students recognized that it might be possible to maintain physical borders with heightened security regimes; cultural borders have become increasingly porous. James Clifford's (1997) idea of 'travelling cultures' captures some of the fuzzy interpenetration of cultural mobility, to which they became increasingly attracted. In Lefebvre's (1991: 38) terms we are speaking of their spatial practices, competence, and performances by which they link together places that are otherwise widely separated. Clifford prefers the word 'travel' to such terms as 'tourism', 'displacement', 'nomadism', 'pilgrimage', and 'migration'. The notion of 'travelling cultures' suggests a two-way relation interactive dynamic process. Terms such as 'displacement' imply a situation of 'bi-focality where an emerging syncretic culture is temporarily separated by erratically enforced frontier controls' (Cohen 1997: 135). The notion of 'travelling cultures' suggests that cultural forms of mobility cannot be bound by exclusionary national security regimes. Imagined constructions of nation-states tie 'locals' to a *single* place, gathering people and integrating ethnic minorities. Global mobility disrupts this logic of national 'belonging', as people are able to imagine belonging to several places at once. Clifford uses the term 'dwelling in travel' to refer to:

> the experiences of mobility and movement, through which people develop a range of new material, spatial practices, that produce knowledges, stories, traditions, comportments, musics, books, diaries and other cultural expressions. (Clifford 1997: 137)

In terms of Lefebvre's (1991: 245–266) theoretical analysis then, we might expect international students to be exemplary carriers of 'travelling cultures', to link countries and allow people, as people, to explore possibilities for post-national politics. International students establish a space of exchange and communication, albeit not necessarily a one-way determinism, but rather a multi-faceted spatial practice. They bring a great deal of cultural information to Australia, couched of course in particular forms or representations. But they also take impressions, images, and information back to their own countries, not only upon their return but also in their regular contact with 'home'. The students were in regular communications with friends and families through telephone, e-mail and voice over internet communications. These contacts suggest that the dislocation and displacement faced by international students have taken particular forms. Their feelings of marginalization and alienation are shaped if not decisively determined by compression of time and space but also, to an extent, by their class position. They remain engaged with developments 'at home', and continue to participate in decision making there. Clifford's phrase, 'dwelling in travel' is indeed an apt one to describe the space within which they negotiate their experiences both in Australia and at 'home'. Upon their return to China, they were more confident about negotiating a cosmopolitan space within which to live.

We should, however, be cautious about generalizing from this postmodernist valorization of mobility and trans-national dwelling. The conditions for a different cosmopolitan identity might be created through the internationalization of higher education. It is wrong to assume that it is an imminent or immediate possibility for everyone, given 'that the space which contains the realized preconditions of another life is the same one as prohibits what those preconditions make possible' (Lefebvre 1991: 189–190). Travelling cultures are not available equally to everyone, and are certainly inflected by gender and class considerations. As bell hooks (1992) has remarked, mobility can often be unsafe and insecure for women. It is experienced differently by (black) women, and its social consequences are unevenly distributed. Under the conditions of globalization, according to Bauman (2000), mobility is available largely to elites. In a challenge to Clifford, he calls them 'tourists', mobile people who contribute in one way or another to the global/local consumer economy. In a globalized society, the mobile tourists—the trans-national businessmen, culture managers, and knowledge workers—are increasingly cosmopolitan. For them and them alone, 'state borders are levelled down, as they are dismantled for the world's commodities, capital and finances' (Bauman 2000: 89). In a significant way, international education is an expression of this consumerist logic. As a commodity, it is now made available internationally to those who have already aligned themselves to the emerging contours of the local/global economy and culture. For them international education has become a site for the spatial production of cosmopolitan identities.

FINAL REMARKS

Our use of Lefebvre's (1991) ideas on spatial production in the field of international education indicates that these international students came to Australian universities with a cosmopolitan imaginary already developed, which their Australian experiences enabled them to develop into a more firmly articulated sense of a cosmopolitan space. While not all the students we interviewed had travelled overseas before, some had. Others had close relatives and friends who had either studied or were currently living overseas. Their exposure to the global media culture had already contributed to their cosmopolitan imaginary. They had thus greater familiarity with Western cultural practices than their universities may recognize. Their international education enabled them to build on this familiarity, and take home with them a sense of belonging to a cosmopolitan space that only their mobility could have provided.

Two further points are worth making here. First, most though not all of these international students had enjoyed considerable class privileges, with their experiences of class increasingly articulating with the degree of their engagement with global economy and culture. In China, international education has become a class marker, to which we need to draw attention when considering how the benefits of trans-national students' mobility are socially distributed. Second, their experiences were unlike those of poor migrants and refugees, for whom there is a major cultural disjuncture between their homeland and their life in the new country. For these international students interactions in Australia were more continuous with their cultural practices at home. The materiality of their class privilege defines the 'scape', to use Arjun Appudarai's (1996) term, within which the students' transcultural contacts in Australia were embedded. This is not to say that the students do not experience marginality. We noted this earlier. However, their marginality had a particular form, from which paradoxically they derived considerable learning. The analysis of spatial perceptions, conceptions, and experiences in these terms foregrounds issues of class, not just mobility but also the prospects for the mobilization of trans-national knowledge workers (Lefebvre 1991: 89).

Aihwa Ong (1999) has also sought to understand these contradictions of local/global mobility. She found that in this era of globalization, mobile individuals develop a flexible notion of citizenship as strategies to accumulate capital and power. The logics of capital accumulation encourage international students such as these to take opportunistic actions in response to shifting political-economic conditions of possibility. In the global arena, powerful incentives exist for such students to emphasize practices that favour flexibility and mobility as they are positioned, but also position themselves in relation to trans-national labour markets, national governments, and the global integration of higher education. Ong illustrates her thesis by referring to evidence of the experiences of migrants from Hong Kong. They were found to demonstrate a flair for prospering in circumstances of political uncertainty and the turmoil of local/global trade. These diasporic Chinese, she suggests, have been the forerunners of the mobile people, who are always 'on the move' both culturally and physically.

This mobility generates a set of trans-national practices and imaginings, resulting in a re-alignment of political, ethnic and personal identities more conducive to plotting the discontinuities between the fixity of politics and the shifts in local/global trade. The market in higher education commodities is constituted and articulated on a global scale via the transportation, communication, and knowledge networks of exchange. The commodification of higher education 'brings in its wake certain attitudes towards space, certain actions on space, and even a certain concept of space' (Lefebvre 1991: 341). This is not to suggest arbitrary geographical and social positionings but rather more creative articulations between subject positions and the regimes of localities, the extended family, the nation-state, and global capital. This places human trans-cultural practices at the centre of discussions of globalization, in the production and negotiation of cultural meanings within the normative framework of late capitalism. And so it is with the experiences of global mobility of students. International education represents a geographical dynamic that helps reshape their identities, cultural tastes, and professional aspirations, but in ways that are neither uniform nor predictable. These students interpreted their experiences in a variety of ways. The cosmopolitan space that they produce was a product of a range of factors, some of which are known to them, others which are not.

ACKNOWLEDGEMENTS

Sincere appreciation is expressed to the international students who gave generously of their time and knowledge to participate in the interviews for this project. This chapter is part of the research project *International Student Mobility and Educational Innovation*, funded by the Australian Research Council.

REFERENCES

Appadurai, A. (1996) *Modernity at large*, Minneapolis: Minnesota University Press.

Bauman, Z. (1998) *Globalization: the human consequences*, Cambridge: Polity Press.

—— (2000) *Liquid modernity*, Cambridge: Polity Press

Berland, J. (1992) 'Angels dancing: cultural technologies and the production of space', in L. Grossberg, C. Nelson and P. Treichler (eds.), *Cultural studies*, New York: Routledge.

Clifford, J. (1997) *Routes: travel and translation in the late twentieth century*, Cambridge, Mass.: Harvard University Press.

Cohen, R. (1997) *Global diasporas: an introduction*, London: UCL Press.

Crang, M. and Thrift, N. (2000) (eds.) *Thinking space*, London: Routledge.

Danaher, P., Danaher, G. and Moriarty, B. (2003) 'Space invaders and pedagogical innovators: regional educational understandings from Australian occupational travellers', *Journal of Research in Rural Education*, 18.3: 164–169.

Elden, S. (2004) 'Between Marx and Heidegger: politics, philosophy and Lefebvre's "The Production of Space"', *Antipode*, 36.1: 86–105.

Hall, S. (1996) *Stuart Hall*, D. Morley and K. Chen (eds.), London: Routledge.

Han, J. (2004) 'Teacher shortages, bilingual teachers and the mobility of transnational knowledge workers', *Melbourne Studies in Education*, 45.2: 99–112.

Hoggart, R. (1958) *The uses of literacy*, London: Penguin.

hooks, b. (1992) *Black looks: race and representation*, Boston, MA.: South End Press

Lefebvre, H. (1991) *The production of space*, Oxford: Blackwell.

Massey, D. (1994) *Space, place and gender*, Cambridge: Polity.

McCarthy, C. (1998) *The uses of culture*, London: Routledge.

Morley, D. and Robins, K. (1995) *Spaces of identity: global media, electronic landscapes and cultural boundaries*, London: Routledge.

Morrow, R. and Torres, C. (2000) 'The state, globalization, and education policy', in N. Burbules and C. Torres (eds.), *Globalization and education: critical perspectives*, New York: Routledge.

Ong, A. (1999) *Flexible citizenships: the cultural logics of transnationality*, Durham, NC: Duke University Press.

Pile, S. and Thrift, N. (1995) 'Mapping the subject', in S. Pile and N. Thrift (eds.), *Mapping the subject: geographies of cultural transformation*, London: Routledge.

Rizvi, F. (2005) 'International education and the production of cosmopolitan identities', in A. Arimoto, F. Huang, and K. Yokoyama (eds.), *Globalization and higher education*, Japan: Research Institute for Higher Education, Hiroshima University.

Robertson, S. Bonal, X. and Dale, R. (2002) 'GATS and the education service industry: the politics of scale and global reterritorialization', *Comparative Education Review*, 46: 472–496.

Singh, M. (2005) 'Enabling trans-national learning communities: policies, pedagogies and politics of educational power', in P. Ninnes and M. Hellsten (eds.), *Internationalizing higher education: critical explorations of pedagogy and policy*, Dordrecht: Springer Verlag.

Singh, M. and Sproats, E. (2005) 'Constructing local/global pedagogies: insights into the learning experiences of international students', *Education and Society*, 23.2: 43–61.

Soja, E.W. (1985) 'The spatiality of social life: towards a transformative retheorisation', in D. Gregory and J. Urry (eds.), *Social relations and spatial structures*, Basingstoke: Macmillan.

—— (2000) *Postmetropolis: critical studies of cities and regions*, Malden, MA: Blackwell.

Tomlinson, J. (1999) *Globalization and culture*, Chicago: University of Chicago Press.

12 Public–private partnerships, digital firms, and the production of a neoliberal education space at the European scale

Susan Robertson

Those following European Union (EU) education policymaking are familiar with the strategic goals for Europe as set out in the 2000 Lisbon Council—*to become the most competitive and dynamic knowledge-based economy in the world capable of sustainable economic growth with more and better jobs and greater social cohesion* (para 5). Observers are also likely to be familiar with the Commission's strategy to generate structural reform and develop its policymaking capacity, as well as the development of actions around lifelong learning and ICT in order to create what is referred to as a European Education Space. These goals, strategies, and actions are, in the words of the Commission, a European response to the challenges posed by globalization, specifically the means to ensure the transition to a digital, knowledge-based economy and society (c.f. European Council 2000; European Commission 2001a).

These initiatives have, for the first time in the history of the EU, placed education high on the political agenda of the Commission, with the specification of the 'Concrete Future Objectives for Education' (European Commission 2001c) and subsequent translations into actions around a series of indicators that benchmark quality (European Commission 2001d). At the same time, the Commission has challenged member states to review and renovate their national education and training systems to deliver '... high educational standards, and to embed a culture of lifelong learning to respond to evolving skills requirements' (European eLearning Summit Taskforce 2001: 2). As Dale (2002) observes, these developments represent '...a very significant sea change in the direction, form and purpose of the relationship between national education systems and the European level'. More particularly, he argues, '...these changes may lead to the development not only of a separate and distinctive agenda at national and European levels, but to a new functional and scalar division of the labour of coordination of these educational agendas' (p. 2) as a new education space is being constructed at the supranational level.

However, this new space of knowledge production at the supranational scale, a means of overcoming blockages as a result of fixed interests at the national scale, is not a neutral process. As I will show in this chapter, scale, and the production of space, is a highly political process and the outcome of particular projects and struggles. Drawing on theories of scale and rescaling (Brenner 1998, 1999, 2004; Smith 1993; Swyngedouw 1992, 1997), I intend to illuminate how education is being reorganized in space, not only horizontally but vertically, with different education activities being organized at different scales. In the case I am looking at, both the European Commission and transnational capital have a vested interest in promoting eLearning and public–private partnerships as a mechanism for producing education in this space. It allows the private sector entry to the European market for digital technologies and the learning market, whilst for the European Commission, the capacity of the private sector is mobilized to ensure the realization of the EU's political project—a competitive European state space.

This chapter begins by outlining the main theoretical arguments around the production of scale and the reworking of state space before turning to the case study: a detailed critical discourse analysis of the report of the European eLearning Summit hosted by the private sector in Belgium in 2001. I show how this initiative is linked to a wider European project—the creation of a European Education Space being strategically developed by the European Commission—arguing that this project can be regarded as a process of territorialization at the European scale. In the final sections of the paper I examine how the interests of the private sector are inserted into the European education scalar project through privileging Public–Private Partnerships (PPPs) as a mechanism for development. I conclude by arguing that this new governance mechanism constructs a set of social relations at the European scale that privileges particular interests; the interests of the transnational digital firms and the kind of knowledge that they are promoting.

THE PRODUCTION OF SCALE AND REWORKING STATE SPACE—THEORETICAL NOTES

Dale's (2002) claim that there is an emerging functional and scalar division in the labour of education is a question begging for a number of reasons. First, it raises the question as to what it means to talk about scale and a division of the labour of education across scales. Dale's thesis is also highly suggestive of a process that is underway; that is, that the labour of education had been previously located in particular sites, and that its redistribution across scales is now part of a new politics of education. Finally, and given that education has been a major project of national states, to talk about the rescaling of the labour of education, in this case to the European level, prompts us to ask questions about the implicit assumptions that we have about the nature of the relationship between existing and emerging scales, and the nature of the institutions, projects, and politics at these different scales.

In order to understand something of the significance of a project like the creation of a 'European Education Space', the term the European Commission uses for this project, and to interrogate the causal logics and possible explanations for this emerging policy trajectory at a supra-national scale, I have found critical theories of space and scale (Harvey 1999; Lefebvre 1991; Smith 1993) and a political economy of scale (Brenner 1998, 2004; Collinge 1999, 2005; Swyngedouw 1992, 1997) to be particularly useful in revealing the spatiality of education projects, and in particular the fact that space and how it is produced matters. Like the other chapters in this book, I am suggesting that this very important dimension has been largely ignored by scholars working on education. In spatializing education in this way, I do not intend to infer that this is a consequence of an emerging and expanding European region. Rather, it is to suggest that education projects are always constituted in space, in turn shaping social relations. It follows, then, that an important task for sociologists of education is to understand the consequences (political, economic and social) of different kinds of education projects as they are spatially organized. And, as Andrew Sayer (2004) reminds us, while space matters as 'all material phenomena necessarily have spatial extension and are spatially located' (p. 267), it does not matter in the same way all of the time. In order to understand the spatiality of education policies, projects, and practices, means '...addressing specific processes, with their distinctive causal powers, situated in specific spatio-temporal contexts' (p. 267). The case study that follows, of the creation of a European Education Space, thus situates the project of inserting a new set of interests and politics into this European space as it is being constructed at a supranational scale.

Several further things should be said before proceeding in order to clarify what I mean. The first concerns the question of scale and its production. As Collinge observes:

> Scale discourse is powerful as it holds out a totalising perspective, seeking to integrate different levels of geographical inquiry. In so doing it draws attention to the division of the global social formation into not only a 'horizontal' structure (in which similar activities are organised at similar scales in different places) but also a 'vertical' structure (in which different activities are organised at different scales covering the same places). (Collinge 2005: 189)

Scale analysis functions by assembling a series of spatial categories into a hierarchical framework that is used to investigate social change. Both Jessop (2004) and Brenner (1998, 2004), for instance, have focused their attention on statehood and processes of rescaling, arguing that diverse areas of national state power, policy formation and socio-political struggles are being redefined in response to both global and local pressures. Further, Collinge's (1999) innovative contribution to scale analysis—'the relativization of scale'—distinguishes dominant and nodal scales in a scalar division of labour. Scale dominance refers to the power which organizations at certain spatial scales, such as the national, are able to exercise over other organizations at higher or lower scales, while nodal scales are defined as scales that are non-dominant in the overall hierarchy of scales. They nonetheless serve as the primary loci for the delivery of certain activities. This is a particularly fruitful set of conceptual innovations for my purposes in seeking to understand the way in which education is being constituted at the supranational or European scale.

The second point concerns the idea of statehood. States have and continue to be important sites for the constitution of education within national territories. Since the 1970s, statehood has been transformed; a consequence of the changing institutional landscape of contemporary capitalism—in particular the mechanisms, sites, and objects of governance. There is now no privileged scale of action as, for example, under the Keynesian Welfare National State (KWNS) regime that dominated the post-war period. Rather, new scales have emerged or existing scales are being reconstituted by new political and governance projects, as, for example, at the local scale through political projects such as decentralization. Activity that was once located at the national scale, for instance aspects of education provision or regulation, is being relocated at different scales, in turn constituting new kinds of social relations within and across scales (Robertson, Bonal and Dale 2002). In other words, the labour of education and its governance increasingly takes place across a range of scales—global to local—while state power is also being dispersed across these scales—delinking it from its concentration at the national scale. This, in turn reveals the methodological nationalist assumptions at work in much thinking about the state and education (see Brenner 2004 on this point in regard to the state in general).

A final note before turning to the case study. Spatial projects are political projects that embrace particular ideas which mobilize particular actors. These processes are, in turn, mediated by competing interests and by the histories of specific contexts. It is thus not possible to determine in advance, or indeed to be able to read off, the precise outcomes of these projects from the public discourses. With these preliminary remarks in mind, let us now turn to a closer look at how these processes are at work in the constitution of a European education space.

THE EUROPEAN ELEARNING SUMMIT AND PPPS

In the Summit Declaration held in La Hulpe, Belgium on 10 and 11 May 2001, the eLearning Summit Taskforce laid out the challenges facing 'Europe' in meeting the goals of the Lisbon Council (2000); to become *the most competitive and knowledge-based economy in the world capable of sustainable economic growth with more and better jobs and greater social cohesion*. The purpose of the Summit was to take forward the European Commission's Action Plan presented by the Council of the European Union in March 2001 to the Member States:

- To develop the comprehensive integration of ICT into education and training
- To create flexible infrastructures that will make eLearning available to all
- To develop universal digital literacy
- To create a culture of lifelong learning
- To develop a high quality European educational content (eLearning Summit Taskforce 2001)

According to the Summit report, 'to meet these goals, Europe needs to expand its educational opportunity'. Educational opportunity is viewed as each individual having access to ICT and a means of developing a level of digital literacy which would enable them to keep pace with economic, social, and technological changes and thus ensure each person will be able to secure their own future through a process of lifelong learning.

The Summit Declaration identified a number of key challenges facing 'Europe'. First, in order to meet the demands of the knowledge economy, there is a need to *accelerate* the process of change and innovation. However, the Summit Taskforce argues that attempts by member states to generate these changes through *bringing teachers alongside* are failing to progress sufficiently rapidly to ensure the realization of the eLearning strategy. The Taskforce observes: 'In many European countries the conditions needed for developing the role of the teacher and enhancing the status of the profession are simply not progressing at a pace that will allow pedagogical innovation to be spread rapidly so that it becomes a systemic part of the education system' (eLearning Summit Taskforce 2001: 2).

Second, according to the eLearning Summit Taskforce (2001), eLearning requires immediate and substantial investments by governments and education and training bodies to generate universal digital literacy. Again, the Taskforce note that the necessary pace of change and the scale of investment in infrastructure, tools, services, and content that will be required, 'must be significantly increased' (p. 2). From there the solution for the Taskforce is an inevitable alternative: that in order to 'provide a *step change* in the implementation of innovative models of eLearning, the transformation of learning institutions, and the social perception of the role and status of educational practitioners' (p. 3, my emphasis), PPPs should be explored as offering considerable potential as a mechanism for delivering education and training into the future given the capital shortages member states face in investing in the public sector. In the view of the Summit Taskforce:

'Active private sector participation in eLearning and ongoing dialogue with the public sector is no longer a viable option but an urgent necessity' (p. 3) and 'a precursor to preparing a *fresh stage in sustainable European cooperation*' (p. 3, my emphasis). The Taskforce Declaration concludes with the recommendation to the Commission that it should 'explore the potential of public private partnerships' (p. 6).

In many respects, the Summit Taskforce report is a remarkable one. To begin, the Summit Taskforce, chaired by IBM Europe, was composed of five companies, IBM, Cisco, Nokia, SanomaWSOY, and Smartforce. It was this group, too, who led the development of the Summit and who have had a significant role in subsequent developments including the development of Career Space, an initiative that I will return to later in the paper. The Summit, hosted at the IBM International Training Centre, attracted over 350 participants from the public and the private sectors, including policymakers from national ministries of education and employment, senior officials from the European Commission, and representatives. At the Summit a further 25 companies (3Com, Apex Interactive, Apple, Auralog, British Telecom, Centra, CEPIUS, Ge.world, Transware, CompTIA, Courseware Factory, De Wilde CBT, Digital Brain, EDS, EdskillsNTO, European Education Partnership, Granada Media, Intel, Interact Group, Manpower, Marconi, Oracle, Sonera), then formed a Steering Group which agreed to take the conclusions and recommendations of the Summit forward. These transnational firms have huge interests in the IT world, including the provision of hardware, software, and education and training. Second, the Summit Report observes that education, as we have known it, must be recast. According to the Taskforce, no longer will education be necessarily delivered via an education system and its teachers. Rather, educators' roles, pedagogical practices, and educational spaces will be redefined within the framework of a partnership between the public and the private sectors.

If one were to read the report of the eLearning Summit Taskforce disconnected from other events and agendas within the EU, it might be easy to conclude that key interests within the private sector had hijacked the agenda of the eLearning Summit and inserted a seemingly tentative claim to the need for (exploring) public–private partnerships as a means for creating capacity. However, as early as 1997, the Commission—in setting out the guidelines for future Community action 2000 to 2006 in *Toward a Europe of Knowledge*, had linked the idea of knowledge and skills for a knowledge economy with the specification of a particular means for bring this about—the private sector. For example, in paragraph three of the Report, under the section *The Parties Involved*—the Commission notes that 'there needs to be a commitment to securing greater involvement of the business sector...the dividing line between the world of education and that of the information society is fluid and connections need to be established in both directions' (European Commission 1997).

The evidence suggests that by the mid-1990s the Commission had a preferred ideological position as to how the European knowledge economy education space should be developed. This ideological preference, of the liberalization of markets and the conditions of trade, was later given structural weight with The Stability Pact, negotiated in Cologne in June 1999 as part of the Commission's conditions for widening the Union to include South Eastern Europe (European Commission 1999). Linked to the Treaty of Maarstricht (1991) with its emphasis on the management of public spending, the Stability Pact (SP) commits EU members and acceding countries to principles of market liberalization (p. 3) while reining in public expenditure:

- public spending must be in surplus—1 to 2 1/2 percent of GDP in 2002;
- central government spending must be in balance;
- overall central government spending must be lower than the growth of the overall budget. (European Council 2000)

While the Stability Pact is invoked as the reason for pressing member states and acceding countries into consideration of private financing of previously publicly funded activity, the Commission had made clear its ideological position and determined that its capacity to construct and govern a European education space in line with the 1991 Treaty of Maastricht and the Commission's economic strategy would require substantial support from the business sector.

EMBEDDING A EUROPEAN EDUCATION SPACE—THE LISBON COUNCIL

Embedding education in a European space is no straightforward process, particularly as ideas like 'European education' are neither pre-existing nor commonsense categories with identifiable activities and institutions, and because the Commission has had limited desire (largely because of its commitment to the principle of subsidiarity) and limited capacity to govern that space in policy terms. However, within the context of creating a competitive European knowledge economy, creating a European education space with a particular mandate and capacity that could be governed was critical. The *Presidency Conclusions* of the European Council in Lisbon, 23–24 March 2000 can be seen as a watershed in this regard (Barcelona European Council 2001). Here the Council specified a clear mandate for education and training, as well as the means for bringing this about. ICT was high on the agenda. According to the Council, investments in ICT infrastructure and digital literacy were critical to developing the services sector and to overcoming the widening skills gap in information technology. The means to do this was along partnership lines.

The Council identifies two kinds of partnership. One kind of partnership would be multi-partners establishing multi-purposed education centres (paragraph 26). This type of partnership would open up the previously closed world of provision to a range of new providers, along with those traditionally in the education sector, as well as being a site that was accessible to different types of (lifelong) learners. A second kind of partnership referred to the means through which the new European education space was to be achieved. Under the heading, *Mobilising the necessary means* (paragraph 41), the Council states that: 'Achieving the new strategic goal will rely primarily on the private sector, as well as on public-private partnerships. It will depend on mobilizing the resources available on the markets as well as on efforts by Member States. The Union's role is to act as a catalyst in this process, by establishing an effective framework for mobilizing all available resources for the transition to the knowledge based economy and by adding its own contribution to this effort under existing Community policies'.

In 2001, the European Commission, in the *European report on the quality of school education* (2001d), laid down a framework for guiding action and mobilizing resources—*16 Quality Indicators* for catalyzing change in the direction envisaged by the Commission. In the area of ICT the key indicator was the 'number of students per computer'—a benchmark according to the Commission that would 'provide an introduction to policy discussion by raising a number of questions about the future place, purpose and practice of ICT in European schools' (p. 7) and 'because ICT is already having a far reaching effect on people's lives and pupils learning, with, for example, 40% of all UK market shares in ICT' (p. 7). While in reality the 'number of students per computer' benchmark of 'quality' tells us little about the conditions of access for pupils in schools, it would seem that its presence as a benchmark is to register the centrality of ICT in the creation of a European education space as part of the European knowledge economy. As Shore (2000) puts it, while Euro-statistics are themselves indices of opinion based upon little more than aggregated data, they:

> are not only powerful political instruments for creating a knowable, quantifiable and hence more tangible and governable 'European population' and 'European space': rather, they are also powerful moulders of consciousness that furnish the meta-classifications within which identities and subjectivities are formed. (Shore 2000: 31)

The European Commission Report also raised questions about the costs/benefits of alternative forms of provision; for example, how much learning can be independent, teacher led, peer group led, or, home school or community based (p. 7). Like previous Commission reports, the European Report also argues: 'The information explosion demands fundamental rethinking of traditional conceptions of knowledge, its transmission, delivery by teachers, and acquisition by students.... All of these areas of knowledge and skills present major challenges to the teaching profession.... Change requires rethinking, reappraisal; re-evaluation of accepted practices, challenging what has always been done and accepted. Change often requires *restructuring and re-culturing of organizations*. It poses new demands on hierarchies, status and relationships' (p. 9, my emphasis).

The Commission Report turns to the difficult question of resources. It suggests that requests for more resources as a typical response and adds: 'more' is not feasible, especially when governments are faced with providing for an increased number of learners in education settings for a longer period of time. In essence, the Report argues that the resource challenge has to be looked at in a different way particularly as 'young people see school structures, curricula and the learning environment' (p. 11) as irrelevant to their lives.

Like previous declarations and reports the Commission identifies the threats to the development of the knowledge economy strategy as lying in both the forces (teachers) and the means (access to computing) of production. That is, teachers lack skills and resist using ICT as they see it as a threat to their jobs (p. 24); the ratio of pupils per computer is still very uneven (p. 52), and in many cases computers in schools are simply not sufficiently up to date to enable them to access programmes that have been developed (p. 53). A preferred solution follows: the flexible knowledge economy means provision should be less institutionalized with individuals assembling their own building blocks of knowledge and qualifications in informal ways and in new contexts (p. 10). The Report observes: 'All member states are realizing that the future brings a monumental challenge to traditional structures of education institutions. This means finding ways of educating people beyond school and outside the classroom, helping them to acquire the skills and competencies that will make them less vulnerable in the global economy' (p. 11). The question then posed is: *How would it be possible to create partnership with institutions or organizations which could help to increase the availability of computers in schools? How can schools be guaranteed a real long-term benefit from such an approach?*

Following the Council Resolution on eLearning in July 2001 (Council of the European Union 2001/C 204/02), an Interim Report—*eLearning: Designing tomorrow's schools*—was released by the European Commission in February 2002 which sought to 'lay the foundations for concrete and sustainable actions' (p. 2) to meet the Commission's knowledge economy goals with an ICT and digital literacy strategy. With enhancing quality and improving access constantly reiterated as the keystones for building the European knowledge society, the report then proceeded to lay out a set of preferred options for what quality and access might mean; flexible and virtual universities, multi-purpose places for learning, the development of ICT Curricula for the 21st Century, and public– private partnerships. Again the issue of resources and the means through which resources might be made available is considered: 'This need is ever more pressing in a more difficult economic environment' (p. 11) while 'Public-Private Partnerships need to be explored' (p. 11). At the same time, it becomes evident at several points in the Report that this European space is more than a 'learning space' in a more traditional sense. While recognizing the recent downturn in the ICT sector and consolidations in the market for e-learning products, the Report observes: 'The global market for eLearning and services is expected to grow strongly in the forthcoming years, providing both a challenge and an opportunity to European education systems and to related economic sectors such as multimedia publishing' (p. 5). The Report concludes by stating: '...it is clear that the eLearning initiative is playing an important role in helping Europe to exploit the use of ICT for education and training, and to realize its potential to be a world leader in learning products and services, and in terms of successfully sharing resources and know-how in education and training' (p. 7).

A EUROPEAN KNOWLEDGE ECONOMY SPACE: SCALE AND THE POLITICS OF TERRITORIALIZATION

Before moving to examine the privileging of public–private partnerships as a means through which the European knowledge economy space is to be realized, I want to consider the social and political processes at work in creating this territory. In particular, and as outlined in the first part of this chapter, the work of the critical geographers and their analysis of space, scale, and territorialization is particularly helpful for understanding this process. So, too, is the work of Shore (2000) and Rosamond (2002) who point out those categories such as 'the European economy' and 'European competitiveness' are not self-evident entities. Rosamond (2002), in an analysis of the construction of a competitive Europe, argues that 'imagining the European economy' is a rhetorical strategy as part of a more complex process of constructing a regime of economic governance at a supranation scale—the European Union. Rosamond shows how 'ideas' like 'competitiveness', can become 'sedimented and "banal" in the sense of becoming commonsensical and barely discussed' (p. 158), in the process constructing identities and subjectivities.

In the same way, we can see how the constant narration of ideas like a 'European education space', a 'competitive and knowledge based economy', and 'public–private partnerships', as well as the institutions engaged with their narration, come to be viewed as commonsense ideas at a scale that sits beyond the national and the local where it is likely that these ideas and political projects will be hotly contested if not institutionally impossible. We can also see the way these ideas are scaffolded into existence and sedimented into institutions and operative networks as material practices through additional policy manoeuvres, such as benchmarking. Finally we can see how these strategies privilege particular kinds of interests and institutional arrangements (as in the eLearning Summit Taskforce and the subsequent development of Career Space) and embed a particular kind of framework for action, a particular type of commonsense. In Robert Cox's (1996: 97) view, a framework for action or historical structure is 'a particular combination of thought patterns, material conditions, and human institutions which has certain coherence among its elements. These structures do not determine people's actions in any mechanical sense but constitute the context of habits, pressures, expectations and constraints within which action takes place'. Crucially, this process is co-constitutive. That is, the construction of space as a particular type of territory, shaped by particular types of ideas, is both the object of and the outcome of struggles between agents that operate at different scales.

The case that I am concerned with, as stated, is the creation of the 'European education space' as a means of becoming a competitive Europe of knowledge and a knowledge economy. This procedure requires engagement in a set of strategic manoeuvres that legitimate the right of a set of supranational institutions (European Commission, Council of Europe, Organisation for Economic and Cooperative Development, European Investment Bank) and trans-national firms operating at a supra-national scale, to create and govern this space. The discursive strategy of the Commission, through these various Reports, is to draw on commonsense discourses of globalization to elaborate upon the external threats, while promoting the uniqueness of the European space. It thus legitimizes policy making in this area and the means of bringing this about. In relation to *threats*, the Commission and key economic actors point to:

1. The inability of national states to generate the level of investment in ICT and education required to keep up to date;
2. The entrenched interests of teachers in national education systems thus making rapid progress difficult; and
3. The difficulties posed by changes in the governance of education systems (devolution) thus limiting the capacity of nation states to direct education systems and ensure equity of access.

With regard to *uniqueness*, the Commission argues that it is only at the European level that the scale and pace of investment is possible. In relation to *means*, the Commission is insistent that the private sector must be involved in the development of education policy and provision. The identification of imperatives means Commission actors are '...then able to offer powerful cases for the development of European-level solutions, delivered through European-level policy instruments and institutions' (Rosamond 2002: 162). Rosamond further observes:

> Such patterns of rhetorical practice are perhaps particular to the Commission and may indeed be part and parcel of the distinctive policymaking dynamics of the EU where supranational entrepreneurs produce analyses of possibilities, ongoing deliberation and interaction. This is especially true of the Commission which, as Thomas Christiansen notes, has developed over time sophisticated strategies for the achievement of its institutional purpose: the expansion of its policy competence. (Rosamond 2002: 162)

Key economic actors, like the large trans-national firms IBM, Cisco, and Nokia, among others, have been actively participating in the creation of a European educational space through generating the conditions for their ongoing and future investment in the lucrative education market without the impediments of existing institutional arrangements, problems of state regulation, and pressure from civil society about the role of large private for profit firms in the education sector. For the EC to foray into education and training on such an unprecedented scale, unhindered by the local and national politics of the member states, it must develop its own system of innovation that enables it to realize a 'quantum shift' in the capacity to bring this about. This means developing the means to go beyond the establishment of objectives and benchmarks. It means drawing upon a set of resources made available through the private sector to provide a particular kind of education that is not dependent upon place but, rather, uses new technologies to operate across boundaries. The construction of this knowledge economy education space carries all the hallmarks of the emerging EU economic space, which, as Rosamond (2002) observes, is quite distinctive: 'It amounts to a quite particular form of economic internationalisation involving the ...freeing of trade and significant deregulation, combined with new rule setting, the development of common policies, the transfer of power to central institutions and the development of redistributive mechanisms' (p. 162). However, the question to be posed is whether the EC has sufficient regulatory apparatus in place, aside from crude measures of quality such as 'the number of computers per pupil', to ensure that social equality is not undermined when the education space is exposed to private for profit interests?

PPPS—INSERTING 'PRIVATE' INTERESTS INTO THE EDUCATION SPACE

I have been tracing the production of a European education space at a supranational scale and the idea that education and training should be delivered in partnership with the private sector. In order to deliver this, the European Investment Bank, a public institution, and a substantial financier of PPP projects since the 1980s in the public sector has prioritized funding education under the Innovation 2010 initiative (i2i) 'in order to close the competitiveness gap between Europe and the US/Japan' (Brown 2004: 3). The i2i program funds three areas: education and training, research and development, and information and communication technologies.

PPPs, as part of a wider competitiveness strategy, have considerable resonance with Third Way or 'Neue Mitte' politics which now characterize many of the restructuring or modernization programmes taking place across Europe, including Germany, Spain, and Greece (see Giddens 2001; Mouzelis 2001). A fundamental idea is that the state should not dominate the provision of public services; rather a range of patterns that include the market and civil society should emerge which enable consumer choice and market competition.

For the European Commission, the idea of partnership has a number of purposes; it enables considerable leverage over a particular type of capacity in the ICT field, the ability to draw upon financial resources/expertise to enhance its territorializing agendas in relation to member states while at the same time suggesting a continuation of the Commission's partnership strategy with member states (see Rodriques and Stoer 2001). In the process, the idea of partnership conceals the extent to which the Commission has sought to be an environment maker rather than an environment taker (Rosamond 2002: 163) in key economic and social policy areas.

However, the insertion of private interests into the European education space is potentially problematic, particularly when we are talking about the introduction of *private-for profit* interests into the partnership. In order to think through this in more detail, it is instructive to look at examples of PPPs. In the EU, the most high profile PPP is the development of 'Career Space' (see http://www.career-space.com). Career Space is a consortium of major information and communications technology companies—BT, Cisco Systems, IBM Europe, Microsoft, Intel, Nokiam Nortel Networks, Philips Semiconductors, Siemens AG, Telefonia S.A and Thales—plus the European Information and Communications Technology Industry Association. Career Space is conceptualized as an alternative site that, it proposes, is critical to helping develop the knowledge economy (it argues that ICT accounts for more than 6.3 percent of GDP). The purpose of Career Space is:

> To develop a framework for students, education institutions and governments that describes the roles, skills and competencies required by the ICT industry in Europe. The first step has been to develop generic skills profiles relevant to key jobs in ICT and to create a dedicated website (www.career-space.com) and use other communication tools to make this information widely available. The generic skills profiles described in this document cover the main job areas for which the ICT industry is experiencing skills shortages. These core profiles describe the jobs, setting out the vision, role and lifestyle associated with them. The specific technology areas and tasks associated with each job are

> also outlined, as well as the level of behavioural and technical skills required to carry out the profiled jobs. (http://www.career-space.com)

The interest of large transnational ICT firms in creating an ICT curriculum framework that gives substance to and which operates in a virtual European education and training space is tied to its own need to generate profits and to shape the conditions that give rise to profitability. The curriculum, however, as Apple (1982) argued more than two decades ago in his work on teachers and texts and the logic of curricular control, is never a neutral project. Rather, the curriculum is a particular set of discourses, shaped by particular sets of ideas about the world and proper social relationships. The creation of a European education space enables the for-profit firms to enter, not only as providers of infrastructures but also as shapers of ideas about the world.

CONCLUDING REMARKS

It is difficult to see how ideas like partnership at the level of Europe can operate with any sense of symmetry of power between the public and the private, especially when it involves powerful companies such as Apple, IBM, and Cisco systems, in short ICT versions of media owner Rupert Murdoch. As Mouzelis (2001: 447) observes in his comments on Gidden's proposed Third Way; a characteristic of late modern society is that economic interests have penetrated the cultural realm and that this process has reached unprecedented dimensions. Mouzelis calls for a regulatory approach that would democratize cultural production by bringing into the sphere of 'civil society' where it would function neither on profit nor state/party logic (2001: 447–449). If, not, he says, we are drifting from a market economy to a market society, a drift that needs to be reversed.

Not only will it be increasingly problematic to reverse these policy tendencies, given the protections afforded private investors under the terms of the WTO/GATS agreements (see Robertson, Bonal and Dale 2002), but it assumes a national focus in its analysis. The real challenge scalar shifts present us with, and which is evident in the case I have been analyzing here, is that these are political manoeuvres by political and economic actors to conceal or reveal particular types of politics. Not only is it difficult to contest what is increasingly viewed as a commonsense solution to the challenges of globalization, but the frameworks which structure the social relations of the European education space are (intentionally) less visible in the political arenas of everyday life. As a consequence, the rescaled functional division of labour and the creation of a European education space are less accessible to challenge.

Because of this, it is particularly important that sociologists of education develop a more complex set of theoretical tools to help understand these changes. This means that it is essential to recognize that 'space matters' in very important ways in education, but quite how and with what consequences, is a matter of specific investigation. Showing how space matters in education reveals many of the implicit assumptions that we have held firm about education; that it is fundamentally a national concern, taking place at the national scale, and that it is the concern of entities we recognize as states—sovereign states. This case study reveals the limitations of this view in important ways. First, that the labour of education while it has largely been located at the national and subnational scales in the post-war period is not *a priori* confined to that scale. Second, that education (or knowledge production and dissemination) can, and is being mobilized as an important project at new scales, such as the supranational. This is increasingly evident as the stakes increase for nations and regions in developing competitive knowledge economy strategies and firms go in search of profits in spaces that are not regulated in the same way or are under construction. Third, that the mobilization of education—under the rhetoric of the knowledge economy—increasingly involves new actors—both non-state and aspirational states, as in the case of the technology firms and the EC. Finally, a spatial analytic in the sociology of education reveals both the poverty of an aspatial approach in showing us what it has concealed, whilst opening up new terrains and tools for exploration of the profoundly political nature of knowledge, its production, distribution, and consumption.

REFERENCES

Apple, M. (1982) 'Curricular form and the logic of technical control: building the possessive individual', in M. Apple (ed.), *Cultural and economic reproduction in education*, London: Routledge Kegan Paul.

Ball, R. Heafey, M. and King, D. (2002) 'The Private Finance Initiative and public sector finance', *Environment and Planning C: Government and Planning*, 20: 57-74.

Barcelona European Council (2001) *Presidency Conclusions*, accessed at http://ue.eu/int/en.info.eurocouncil/index.htm

Brenner, N. (1998) 'Between fixity and motion: accumulation, territorial organisation and the historical geography of spatial scale', *Environment and Planning D: Society and Space*, 16. 1: 459–481.

——— (1999) 'Globalisation as reterritorialisation: the rescaling of urban governance in the European Union', *Urban Studies*, 36.3: 431–451.

——— (2004) *New state spaces: urban governance and the rescaling of statehood*, Oxford: Oxford University Press.

Brown, T. (2004) Presentation by Director General for Lending Operations in Europe, European Investment Bank to 10th EIB Forum, Warsaw, 14–15 October, accessed at http://www.eib.org/Attachments/general/events/forum2004_26_brown.pdf

Collinge, C. (1999) 'Self-organisation of society by scale: a spatial reworking of regulation theory', *Environment and Planning D: Society and Space*, 17: 1–23.

—— (2005) 'The *différance* between society and space: nested scales and the returns of spatial fetishism', *Environment and Planning D: Society and Space*, 23: 189–206.
Commission on Public Private Partnerships (2001) *Building better partnerships: a final report*, London: Institute for Public Policy Research.
Council of the European Union, Council Resolution July 2001 (2001/C 204/02).
Cox, R. (1996) *Approaches to world order*, Cambridge: Cambridge University Press.
Dale, R. (2002) *The construction of a European education space and education policy*, A paper presented to the European Science Foundation—Exploratory Workshop, 'Globalisation, Education Restructuring and Social Cohesion in Europe, Barcelona, Spain, 3–5 October.
European Commission (1996) *Learning in the information society, 1996–1998*, accessed at http://europa.eu.int/comm/education/elearning/doc_en.html
—— (1997) *Toward a Europe of knowledge*, accessed at http://Europe.eu.int/comm./education/orient/orie-en.html
—— (2000) *Toward a Europe of knowledge*, 2000–6.
—— (2001a) *Designing tomorrow's education*, An Interim Report, Commission Staff Working Paper, accessed at http://europa.eu.int/comm/education/elearning/doc_en.html
—— (2001b) *Designing tomorrow's education*, accessed at http://europa.eu.int/comm/education/elearning/doc_en.html
—— (2001c) *The Concrete future objectives of education systems*, accessed at http://europa.eu.int/comm/education/objet_en.pdf
—— (2001d) *European report on the quality of school education–16 Quality Indicators*, accessed at http://europa.eu.int/comm/education/life/15indicators_en.pdf]
European Council (1991) *The Treaty of Maastricht*, accessed at http://www.uni-annheim.de/users/ddz/edz/doku/vertrag/engl/m_engl.html
—— (1999) *The stability pact for South Eastern Europe*, Cologne, June.
—— (2000) *Update of the Luxembourg stability programme for 1999–2003*, accessed at http://www/etat.lu/FI/
European eLearning Summit Taskforce (2001) *Summit declaration*, accessed at http://europa.eu.int/comm/education/elearning/summit.pdf
Giddens A. (ed.) (2001) *The global third way debate*, Cambridge: Polity.
Harvey, D. (1999) *The limits to capital* (2nd ed.), London: Verso.
Heyneman, S. (2001) 'The growing international commercial market for educational goods and services', *International Journal of Educational Development*, 21: 345–359.
International Cooperation Europe Ltd. (2001) *Curriculum development guidelines: new ICT curricula for the 21st century: designing tomorrow's education* (http://www.career-space.com)
Jessop, B. (2004) *From localities via the spatial turn to spatio-temporal fixes: a strategic relational odyssey*, SECONS Discussion Forum, Contribution No. 6, University of Bonn.
Lefebvre, H. (1991) *The production of space*, Oxford: Blackwell.
Mouzelis, N. (2001) 'Reflexive modernization and the third way: the impasses of Gidden's social democratic politics', *The Sociological Review*, 49.3: 436–456.
Robertson, S. Bonal, X. and Dale, R. (2002) 'GATS and the education service industry: the politics of scale and global re-territorialisation', *Comparative Education Review*, 46.4: 472–496.
Rodrigues, F. and Stoer, S. (2001) 'Partnership and local development in Portugal: from "globalised localism" to a new form of collective action', in M. Geddes and J. Benington (eds.), *Local partnerships and social exclusion in the European Union: new forms of local social governance*, London: Routledge.

Rosamond, B. (2002) 'Imagining the European Economy: "competitiveness" and the social construction of "Europe" as an economic space', *New Political Economy*, 7.2: 157–177.

Sayer, A. (2004) 'Seeking the geographies of power', *Economy and Society*, 33.2: 255–270.

Shore, C. (2000) *Building Europe: the cultural politics of European integration*, London: Routledge.

Smith, N. (1993) 'Homeless/global: scaling places', in J. Bird et al. (eds.), *Mapping the futures: local cultures, global change*, New York: Routledge.

Swyngedouw, E. (1992) 'The mammon quest: globalisation, interspatial competition and monetary order: the construction of new spatial scales', in M. Dunford and G. Kafkalas (eds.), *Cities and regions in the New Europe*, London: Belhaven Press.

—— (1997) 'Neither global nor local: globalisation and the politics of scale', in K. Cox (ed.), *Spaces of globalisation: asserting the power of the local*, London: Guildford Press.

13 Deparochializing the study of education

Globalization and the research imagination[1]

Bob Lingard

Industrial capitalism and that of services, colonial systems, modernity, and the information revolution have been managed by the West, and this has implied the generalization of Western culture—not so much as an ethnic culture but as a 'metaculture'. However, the objectives of conversion and domination of this metaculture have also necessarily opened its information circuits to wider access, making it in turn vulnerable to exploitation by the 'Other' for its own, different ends, transforming the metaculture from within and without. Western metaculture has paradoxically become a medium for the affirmation and globalization of differences beyond their local referents (Fisher and Mosquera 2004: 5).

Today we live and work in globalized spaces and places and need to recognize this in our theorizing, research, and pedagogies within the academy. This means we need to challenge any taken-for-granted and easy assumption of a society/nation homology in research and theorizing. Today society is in some ways simultaneously local, national, regional, and global in terms of effects, experience, politics, and imaginaries. Further, these spaces are imbricated in unequal power relations, which reflect both contemporary geo-politics and past political struggles. Residual, dominant, and emergent geographies of power, including those of the colonial past and postcolonial present, are at play across this global space and made manifest in vernacular ways in the local and the national. Western 'metaculture' relates to these cartographies and geographies of power and includes academic theories, epistemologies, and research methodologies. Recognition of theoretician, researcher, and teacher positionality within Western universities and their relationships to these cartographies of power is an important beginning for challenging the silent valorization of Western epistemologies. Such a challenge is also imperative for moving towards what Arjun Appadurai (2001) calls the deparochialization of research and a strong internationalization of the Western academy, in the light of enhanced global flows of students and academics as part of the ethnoscapes, mobilities, and networks of globalization.

This chapter utilises postcolonial theory, including the work of Gayatri Spivak, Edward Said, Linda Tuhiwai Smith, and Arjun Appadurai, and Pierre Bourdieu's work to draw attention to these usually overlooked matters in respect of policy studies in education and teaching educational policy studies to travelling non-Western students studying within a Western university. The perhaps contentious conjoining of Bourdieu and postcolonial theory, is justified in terms of Bourdieu's (2003) recognition that contemporary globalization read performatively as neoliberal politics and economics has been a political project with a history and that the effects of globalization within nations are in some ways contingent upon the extent of 'national capital' held and developed historically. This seems to appositely complement the postcolonial recognition of the colonial past forever haunting the postcolonial present and the politics involved in challenges from postcolonial aspirations to these resistant residues in an attempt to strengthen 'national capital' for a more autonomous postcolonial politics. There are also evident complementarities between Bourdieu's rejection of an epistemological 'state of perfect innocence' (Bourdieu 1999b: 608) and Appadurai's (2001) postcolonial call for 'epistemological diffidence', both of which recognize the constructedness of knowledge and thus potential disfiguring through failure to acknowledge this.

The chapter specifically works in dialogue with Appadurai's paper, 'Grassroots Globalization and the Research Imagination' (2001) in an attempt to outline some necessary changes in researching education. The widening of the object of study of educational research, with the breakout of pedagogies from specialist educational institutions and the globalization of the spatial location of educational policy effects, challenges the institutional and national focus of educational research. Further, the new multi-modal and multi-directional circuits of power associated with globalization, and related flows of students and researchers, challenge many of our theoretical, epistemological, and methodological assumptions about doing educational research. The educational research imagination needs to be exercised to meet these challenges of globalization.

In his paper, Appadurai suggests the need to deconstruct in both an anthropological and pragmatic sense, the 'taken for granted' of the contemporary system of research in the context of globalization and its flows of capital, people, ideas, images and technologies, and disjunctions and related asymmetries of power. Specifically, he calls for a 'deparochialization of the research ethic' (Appadurai 2001: 15) and a strong internationalization of research. This is necessary, he avers, to the globalization of the knowledge of globalization and for challenging globalization as simply neoliberal economics.

The argument of this chapter is that deparochialization and strong internationalization are also required of educational research. It provides two narratives regarding the deparochialization of research in education. In narrative one, following my own educational policy research trajectory from a national policy focus to researching the effects of the global in national and state educational policy developments in Australia, an argument will be sustained that a new imaginary is imperative because factors affecting national systems of education, educational policy, and local pedagogies, now extend well beyond the nation. This is part of the larger project which suggests that the mobilities associated with globalization demand a rethinking of implicit taken-for-granteds in social theory of the relationship between theory and society as a nationally bounded space (Beck 2000; Urry 2000, 2002). The work of Bourdieu (1996, 2003) on fields will be utilised to suggest an emergent global educational policy field with the extent of national autonomy mediated by the strength of specific national capitals.

Narrative two of deparochialization draws on my experiences teaching and directing a doctoral programme in the Caribbean for the University of Sheffield, and teaching about globalization and educational policy on a Sheffield-based master's programme in educational policy and research, where most of the students are from China. These experiences made me aware of those elements that need to be considered and traversed as part of Appadurai's (2001) desired move towards a stronger definition of what the internationalization of research in education might entail.

Appadurai (2001) suggests a number of ways in which the research ethic might be challenged; he argues for a reconnection with earlier pre-research paradigm thinking premised on a strong moral position, reconnection with the style of argumentation of public intellectuals, and consideration of research linked to policy making and state functions in a range of nations, particularly the developing world. Most Sheffield doctoral candidates in the Caribbean argue the individual benefits of their research. However, they also argue a contribution to knowledge *raison d'etre*, but importantly for the thesis here, they argue most strongly about the building of research capacity in the Caribbean. They desire to contribute to research informed politics and policy making as a speaking back to the requirements and conditionalities of international donor agencies such as the World Bank and informing policy making within the local state. In their view, stronger national research capacity would offer the possibility for stronger national mediation of the effects of such donor agencies on educational policy agendas. Indeed, they recognize that the capacity to *do* research rather than simply *be* researched is central to postcolonial aspirations. However, such capacity and practices also evoke potential and significant challenges to theory and methodologies, which are largely western and metropolitan in origins.

DEPAROCHIALIZING RESEARCH

Appadurai postulates that 'epistemological diffidence' is necessary to the project of deparochialization of research, and suggests that this can be contrasted with the epistemological certainty of dominant forms of modernization theory of the 1950s and 1960s and its effects, particularly in what was then called the 'Third World'. That theory accepted without question that theory and research were metropolitan, modern, and western, while the rest of the world—in this case the developing world, now the global south—was simply a research site to test and confirm such theory. Here relations of researcher and researched paralleled relations of colonizer and colonized, even within decolonizing and postcolonial politics and aspirations. As an aside, as a university student in Australia in the 1960s and 1970s, I had similar experiences that theory was something Australian academics and researchers utilised and applied, but did not develop. The same dependency is evident in the contemporary Australian academe, when international recognition is utilised to evaluate research output; here international sometimes is read as non-Australian but English speaking, a very different reading from that in the United Kingdom or United States of America, where it is read as high quality with large impact, while potentially remaining deeply parochial.[2]

On this politics of research and representation, Linda Tuhiwai Smith (1999) opens her *Decolonizing Methodologies*, a study of the relationship between research and Indigenous peoples and knowledges in New Zealand, with the statement: 'From the vantage point of the colonized, a position from which I write, and choose to privilege, the term "research" is inextricably linked to European imperialism and colonialism' (Smith 1999: 1). This is Said's (2003) point that even the most arcane theories, including I would note research methodologies, are distorted to some extent by relations of power and politics in both macro and micro relations. Postcolonial insights are important to the decolonizing and deparochializing strategies required.

Appadurai's call for the deparochialization of research is set against the effects of globalization after the Cold War and the presence of one global superpower operating unilaterally. An important element of this context is the enhanced flows of students in the global market of higher education and the attraction of the US as the place for graduate study for potential students around the globe, including students from China and the global south. Here there is the dangerous possibility for globalization to be synonymous with Americanization and a weak form of internationalization of research to occur, which simply means the inclusion of more people in the conversation of researchers without challenging its western, metropolitan form. The disjunctions between the various human and cultural flows of globalization, however, also allow for some optimism for a stronger form of internationalization than this. There are other regional student flows, for example, within South East Asia to Singapore and within the Caribbean to Cuba, which offer some challenge to the allure of the west in the global student market.

In rejecting a stance of epistemological innocence (Bourdieu 1999b), Appadurai suggests that some 'epistemological diffidence', and what Tuhiwa Smith (1999) would see as 'epistemological openness', can help in a strong internationalization of research, and I would add theory, and thus move us beyond what we might see, *à la* Said, as an Orientalist approach to both research and theory. Such a project, according to Appadurai, needs to be aligned with 'grassroots globalization' or 'globalization from below'. This is the question of whose globalization and offers a challenge to globalization from above (in an economic policy sense in particular) as driven by some international organizations and US foreign policy and cultural industries. In *Modernity at Large*, Appadurai (1996) spoke of 'vernacular globalization', the ways in which local sites and their histories, cultures, politics, and pedagogies mediated to greater or lesser extent the effects of top-down globalization.[3] This is the outcome of relations and tensions between the context productive and context generative effects of globalization; some local sites are more able to be context generative and mediate global effects. The idea of globalization from below is about strengthening this mediation and enhancing global connections across NGOs and other groups of resisters to globalization from above, read simply as neoliberal economics, what Bourdieu (2003) calls the performative construction of globalization. Globalization from below could also be seen to be aligned with a postcolonial aspirational politics.[4]

Gayatri Spivak (2003) has made some telling comments about the disjuncture at times between the local, educated elites in the global south, who in human rights discourses are the 'righters of wrongs', and the rural poor within their nations. The former probably have more in common with the educated elites of the global north. These are the epistemological disjunctures and exclusions of globalization about which Appadurai (2001) speaks. As Spivak argues, the sort of education required in the south is not that which would 'make the rural poor capable of drafting NGO grant proposals' (Spivak 2003: 173–174)! Importantly, echoing Said's observation about the elite universities in the US being the last utopias for free and questioning thought, she argues, rightly in my view, that globalization from below also requires an unmooring of such education from its elite location in both the North and the South. This would go well beyond Freire's banking conception of education towards conscientization, and 'beyond literacy and numeracy and find[ing] a home in an expanded definition of a humanities to come' (Spivak 2003: 173).

As noted already, Appadurai writes about the need for research to examine its own 'taken for granteds' in the present globalizing context. These include 'systematicity, prior citational contexts, and specialized modes of inquiry', replicability, along with 'an imagined world of specialized professional readers and researchers' (Appadurai 2001: 12), which taken together work to inhibit the deparochialization of research, its theories and methodologies. The argument here is that a particular postcolonial politics is a useful starting point for a re-reading, re-examining, re-imagining, indeed deparochializing of re-search in the globalized context of American and western neo-colonialism, a reality often glossed over in talk about globalization.

DEPAROCHIALIZATION OF THE FOCUS OF EDUCATIONAL POLICY STUDIES: NARRATIVE ONE

In my own educational policy research in Australia, I have shifted scalar focus from the national to the global. During the late 1980s and early 1990s under Labor governments, there was a move towards national policies in schooling, despite the federal political structure and the states being responsible for schooling. I researched that move through in-depth interviews with the policy and political elites and became aware of the emergence of global pressures on national policy and of what Bourdieu (1995) might have called an emergent 'world or global educational policy field' (Henry et al. 2001; Lingard 2000; Lingard et al. 2005; Ozga and Lingard 2006; Taylor et al. 1997). This made me aware of what I now would describe as the need to deparochialize educational policy studies—the focus of this narrative.

The reality of trans- and supra-national processes labelled as globalization has challenged contemporary social theory. In sociology, Urry (2000, 2002) has argued the need to refocus from the social as society to the social as mobilities, indicating the weakened connectivity between society and nation-state and the stretching of networks across the globe. The spatial turn in social theory has been another response. Massey (1994: 2) has noted that the 'spatial is social relations stretched out'. Castells (2000) argues that today power is located in flows, while most people still live in the space of places; this disjunction, he suggests, results in political schizophrenia. Other theorists have demonstrated the implicit national space of much social theory and a complementary 'methodological nationalism' (Beck 2000). Bourdieu, for example, has observed how '[i]ntellectual life, like all other social spaces, is a home to nationalism and imperialism' (1999b: 220) and that 'a truly scientific internationalism' requires a concerted political project; this is another way of expressing the project in which Appadurai invites us to participate.

Bourdieu's theoretical stance and methodological disposition allow a way beyond such spatial and national constraints, a necessary position for analyzing and understanding global effects in contemporary educational policy and the emergence of a global policy field in education. Globalization has resulted in the compression of time and space, which has had the phenomenological effect of enhanced awareness among (privileged) peoples across the globe of the world as one place, evidenced in, for example, talk of the 'world economy', 'world recession', 'global warming', 'world heritage sites', 'world policy', 'global educational indicators', and so on. Appadurai (1996) speaks of the flows and networks across the globe that render national boundaries more porous. Focusing on the various cultural flows associated with globalization, he also speaks of ethnoscapes, mediascapes, technoscapes, financescapes, and ideoscapes (p. 33ff), which are related to the emergence of the postnational era. Castells (2000) also argues that society is now organised around flows, namely, 'flows of capital, flows of information, flows of technology, flows of organisational interaction, flows of images, sounds and symbols' (p. 442), with technology facilitating these flows via nubs and nodes located across the globe that are dominated by elites of various kinds. However, the postnational accounts proffered by Appadurai and Castells probably overstate the 'porousness' of national boundaries, and the extent of the postnationality of the present, particularly since September 11, the bombings in Madrid and London, where such boundaries have become somewhat less porous in the so-called 'war on terror' (Gregory 2004; Rizvi 2004).

The relational approach and that of fields as a social space, rather than being a specific material, grounded space, also allows for the stretching of the concept of field in an elastic way across the social space of the globe, taking us beyond empirical investigation of only the local and the national to the nested regional, international, transnational, and global spaces of educational policy production. To paraphrase Massey (1994), the global educational policy field sees the social relations of educational policy production stretched out.

Much writing about globalization, as Rizvi (forthcoming) argues, has reified the concept, failing to historicize it and to recognize its hegemonic role, while neglecting the asymmetries of power between nations and colonial and neo-colonial histories, which see differential national effects of neoliberal globalization. It is the argument here that Bourdieu's work potentially offers a way beyond such reification of globalization and allows for an empirically grounded account of the constitution of a global policy field in education, an example of globalization from above. In his later, more political work Bourdieu (1998, 1999a, 2003) is concerned with the politics of globalization, read mainly as the dominance of a neoliberal approach to the economy. Interestingly, this appears to argue against his earlier work on the various fields of social and cultural production, which suggested a relative autonomy of the logics of practice of each field. Instead, he seems to be suggesting ways in which global neoliberal politics have dented somewhat the relative autonomy of the logics of practice of many social fields, including that of the educational policy field, which has become more heteronomous as a sub-set of economic policy. The media field and its logics have also affected the degree of autonomy of educational policy production (Lingard and Rawolle 2004).

While much talk about globalization has reified the concept, failing to locate agency in respect of its workings, Bourdieu's approach allows for an empirical investigation of the constitution of the global economic market. There is, however, another potential pitfall here, that of over-emphasizing collective agency in the process. This is why empirical investigation of these processes is needed. And of course there are the counter-hegemonic effects of anti-globalization movements and globalization from below.

In a homologous fashion, I would argue that the global field of educational policy is also a political project and yet another manifestation of the emergent politics in the age of flows and diasporas of people and ideas across the (more or less) porous boundaries of nation states in both embodied and cyber forms. Political imagination has been a central component in the creation of the global market and the global field of educational policy, just as political imagination is necessary to challenge their effects. Different nations, now located in a post Cold War world with one superpower seemingly committed to unilateralism are, of course, positioned differently in terms of power and the strength of national capital within these global fields of economy, governance, and educational policy. The differing 'reference societies' (Schriewer and Martinez 2004) for different national educational systems within the global educational policy space also tell us something about political history and contemporary global politics. Policy making in education within the nation and at sub-national and local levels, nonetheless, still remains important.

Drawing on Bourdieu (2003: 91), it can be argued that the amount of 'national capital' possessed by a given nation within these global fields is a determining factor in the spaces of resistance and degree of autonomy for policy development within the nation. Here the global south is positioned very differently in relation to the educational policy effects of the World Bank and other international agencies from the global north. National capital can be seen to mediate the extent to which nations are able to be context generative in respect of the global field. Under globalization and the emergence of global fields, the sovereignty of the different nation states is affected in different ways; as Jayasuriya (2001: 444) suggests, 'the focus should not be on the content or degree of sovereignty that the state possesses but the form that it assumes in a global economy'. Further, as an example of Appadurai's (1996) point about the disjunctions between the cultural flows of globalization, national capital in respect of the global educational policy space is not exactly homologous with economic capital. The iconic status of Singapore and Finland in this global educational policy space in terms of Trends in International Mathematics and Science Study (TIMSS) and Programme for International Student Assessment (PISA) results respectively, is a good illustrative case in point and also a good example of the elision of contextual differences within this global educational space. Numbers are central to the decontextualized character of the global educational policy field.

In research on the OECD, for example, it was shown how OECD policy work had more salience in the policy culture of the more peripheral Scandinavian countries and Australia than in Britain, while it was demonstrated how the USA played a hegemonic role in the creation of the educational indicators project which was created against the policy inclinations of some OECD insiders (Henry et al. 2001). In work on the emergence of a European educational policy space, it was shown how a form of structural adjustment saw convergence effects in the educational policies of net benefactor countries within the EU, that is, in the least powerful and least developed countries of the Mediterranean rim such as Portugal and Greece, with talk of a supranational 'magistrature of influence' (Lawn and Lingard 2001). That research also demonstrated the awareness of policy makers of this postnational policy and the lack of recognition of it by school-based practitioners. In the developing countries of the world, the effects of the World Bank upon educational policy are palpable, though not necessarily taken up unconditionally, while not recognized or felt in Europe, North America or Australia.

Politics can also resist the global dominance of neoliberalism as suggested by Bourdieu (2002) in his article on 'The politics of globalization' which *Le Monde* published on the day after his death, and in which he observed that 'the construction of a Social Europe' would be a good bulwark for resistance against 'the dominant forces of our time'. The state is not powerless in the face of globalization, but different states have varying capacities and varying amounts of 'national capital' to manage their 'national interests'. It is this 'national capital' that ought to be the focus of research on educational policy.

When combined with the argument about competing logics of practice, this offers another way of thinking about policy/implementation relationships across the contexts of policy text production and that of practice, both across and within national borders. The educational policy field today is multilayered stretching from the local to the global. In terms of this multilayering, Mann (2000) speaks of five socio-spatial networks, namely local, national, international (relations between nations), transnational (pass through national boundaries), and global, which cover the globe as a whole. Theorization about and empirical investigation of educational policy fields today must recognize the growing global character of relations between national policy fields and international fields, re-emphasizing Bourdieu's conceptualization of fields as social rather than purely geographical spaces. Such an account needs to see the various networks, referred to by Mann as sitting within a global educational policy field.

The educational policy field thus now demands an empirical and theoretical stretching beyond the nation. The argument here is that Bourdieu's approach enables us to do this in both theoretical and methodological senses. The model that I advance, of adapting Bourdieu's notion of field to examine the relations between global, international and national educational policy fields, offers a different way to locate the practices and products of policy. This global educational policy field encompasses the contexts of the policy cycle of which Ball (1994) writes, and offers some analytic gains in locating the effects of particular policies. That is, it caters for these matters and also offers a particular way of utilizing Bourdieu's concept of field to discuss issues around the impact of different fields on one another within national fields of power, and of different scalar levels of fields also affecting one another. All three contexts of the policy cycle, the context of policy text production, the context of influence and the context of practice are affected in different ways by globalization (Ball 1998), through both its policy mediation and more direct effects.

Bourdieu's conceptual apparatus allows for a deparochialization of educational policy studies by setting a frame for the theoretical and empirical study of the world educational policy field. What has been provided in this narrative is a skeletal outline of how and why the scalar focus of educational policy studies must be stretched and the contribution that Bourdieu's approach can make to that deparochializing project. What is needed now is much empirical investigation.

DEPAROCHIALIZATION OF THEORY AND METHODOLOGY IN EDUCATIONAL RESEARCH: NARRATIVE TWO

> I've been teaching now for almost forty years. And I've always learnt during the actual class. There's something that eludes me when I read and think without the presence of students. So I've always thought of my classes as not a routine to go through but rather an experience of investigation and discovery. (Edward Said in Viswanathan 2001: 280)

My pedagogical experiences articulated below align with Said's experience of learning through teaching. These two somewhat cryptic narratives are provided to raise issues in relation to the project of strong internationalization of research and the globalization of knowledge. My positionality here is as an Australian teaching in an English university (now Scottish), myself part of the flows of globalization, and teaching international students located in Sheffield and students of the University of Sheffield located in the Caribbean.

Chinese international students at the University of Sheffield

I have worked with three cohorts of full-time MA students in educational policy and research at the University of Sheffield. The majority of these students in each year of the programme have been young women from China, self-sponsored, full-fee paying, who have been teachers of English, mostly in universities, but some in schools, in various parts of China. One Chinese male I have taught and supervised was in the international office of a large and important Chinese university. His research was about the meanings of internationalization for Chinese universities, but all framed by how, given language difficulties, China generally and his university in particular, could attract some of the full fee-paying students who now moved in flows towards the US (Marginson 2004). He would joke that when China became a world superpower then the flows of international students might be greater in China's direction, indicating the link between economic power, knowledge, and language.

What struck me about these students in teaching globalization and educational policy, as well as supervising dissertations was their positive reading of globalization. Unlike much of the literature on globalization, which sees the dominant reading as that of neoliberal economics, with all its attendant negative effects within and across nations, these international students regard globalization as a positive phenomenon. We become aware here of the multifaceted nature of globalization and the significance of positionality to reading it. In this respect, it is interesting to contemplate Tikly's observation about globalization and education theory:

> It has been a shortcoming of much of the existing literature on globalization and education that the specific contexts to which the theory is assumed to be applicable have not been specified. It is problematic to assume that there is one superior vantage point from which global forces can best be understood. (Tikly 2001:152)

Deng Xiaoping is seen as the centrally important figure for the Chinese students in relation to globalization, opening up China to the world in various ways after its inwardness during the period of the Cultural Revolution. These students are, of course, the beneficiaries of globalization; they are part of the ethnoscapes of globalization. They also make us aware of the centrality of English, along with ICTs, as the bearers of globalization and invite us to think of how a common language produces common ways of seeing. The question of language needs to be a central one in the attempt to deparochialize research (see also Mufti 2005, in respect of English and the study of comparative literatures).

For some of these students, globalization is linked to aspirations for a more open political system, and their own desire for participation in global labour markets. They see the possibilities for pressures from globalization and demands and contradictions, which stem from Chinese 'market socialism', for making things better. This is the case in particular with their still somewhat muted criticisms of the dual administrative structures within their own state controlled work organisations of professionals and party officials. They see this as meaning their organisations are still sclerotic when it comes to change, but global pressures and the embrace of the market, will mean they will *have* to change, and this will be beneficial. At the same time as desiring a new style, more open and pluralistic polity, they fear a disastrous Soviet type transition from command economy to market capitalism. They desire a new political way and see globalization as providing an aegis for this, while profoundly recognising that neoliberal globalization manifests very differently within market socialism.

The question of criticality in content and pedagogy about globalization becomes very interesting. The students appear able to be critical to a point, within certain parameters: always critical of the Bush regime, for example. In terms of the research topics the students have pursued, these are in many ways accounts of their own experiences, as beneficiaries of one of the central flows associated with globalization, but utilizing western research methodologies and theories. In terms of the citational contexts in which their research is located, it tends to be very limited, or about similar issues in other contexts, not their own. This is certainly the case in respect of methodological issues as well. There is also the interesting issue of translation in respect of much of their interview data and all the complexities that entails. Further, they are resistant to the use of the personal in their writing, and perhaps research and write within an academic tradition of smaller citational frames and where researchers take research problems in terms of the infidelities of the implementation of government determined policy.

Teaching and supervising Caribbean Students

I was Chair of the University of Sheffield's Caribbean programme, which has been functioning since 1989. The political frame for this programme has been postcolonial, taking the local and national civic missions of the University onto a global and postcolonial stage. However, I also recognize the old colonial relation as backdrop to the involvement of an English university in the Caribbean, even though the programme is seen more as part of the globalization of the University's civic mission, rather than as part of a revenue raising market.

While recognising the multi-meanings of 'post' when attached to colonialism (Shohat 2000)—after, beyond, neo, building upon, aspirational, etc., and indeed its ambivalence (Radhakrishnan 2000), the programme has been framed by a reading of post as aspirational, in terms of the pedagogies, methodologies, the topics, and intellectual resources utilised in the teaching and research and in terms of the desired outcomes. In this respect it is significant that the programme works with Caribbean-based tutors and works through a local Caribbean consultancy firm. The programme's underpinning aspirational definition recognizes the differences between postcolonialism and postmodernism, particularly in relation to provisional commitments to the Enlightenment project in respect of equality and democracy. As Said, in the 1995 Afterword of *Orientalism*, puts it:

> Yet whereas post-modernism in one of its famous programmatic statements (by Jean-Francois Lyotard) stresses the disappearance of grand narratives of emancipation and enlightenment, the emphasis behind much of the work done by the first generation of post-colonial artists and scholars is exactly the opposite: the grand narratives remain, even though their implementation and realization are at present in abeyance, deferred, or circumvented. (Said 2003: 351)

Specifically in the Caribbean, I have worked with master's students in their course work and in dissertation supervision. The thing I notice is the impact of Freire's (1972) 'banking education' on these students. Criticality and the capacity and right to express one's own voice have not been allowed. In early writings, the student's voice is usually missing. Quotations from others sometimes substitute for an argument. On querying this, many of the students say they were not allowed to express their argument and voice in their undergraduate courses in the Caribbean. Often the literature reviews, which the students prepare as part of their dissertation work, deal only with US literature without a mention of its specificity or more accurately, its specificity segued as universal. The presence of one citational context has effects, as does the absence of another, that is, an extensive literature about their research topics in the Caribbean. These research students are out of place, as it were. Further, in the early stages of the programme many students have a narrow conception of what research in the social sciences ought to look like. Their dependency disallows experimentation and creativity. It is reflection upon these matters, through what we might tentatively call 'postcolonial pedagogies' and 'conscientization', that all of us working in the Caribbean programme seek to encourage a disposition of criticality and a postcolonial attitude. Incidentally, Dimitriadis and McCarthy (2001) provide an especially valuable account of postcolonial pedagogies that are useful in the west. Postcolonial histories of education in the Caribbean are provided as a contribution towards strengthening public memory of colonialism and its residual legacies as a way of challenging the amnesia resulting from the colonial experience. This postcolonial pedagogy works against the silencing of the local and of the past; such silencing and forgetfulness are effects of colonialism. Said (2001: 502) describes this desired disposition as: 'a sense of critical awareness, a sense of scepticism that you don't take what's given to you uncritically'.

My work with Caribbean doctoral students has been of a different order and I am speaking here of the professional doctoral programme taught in the Caribbean, as well as supervision of Caribbean PhD students in Sheffield. Teaching about educational policy to the doctoral students in the Caribbean, I am always struck by their deep awareness of their political location in the contemporary globalized world. Of all the educational policy students I have taught, they are the ones who immediately raise issues to do with globalization, global geo-politics, the World Bank, and other donor agencies and their effects on educational policy production, the globalized policy discourses in education, which they ventriloquize, and their very real effects on policy implementation and pedagogies. They are aware of a global educational policy field, the focus of the first narrative in this chapter. They articulate the negative effects of running to policy tunes set elsewhere in terms of timing of policy agendas and implementation. They talk about the effects of WTO decisions on the banana industry in the Caribbean; they know about the creation of new regional blocs and markets in the Caribbean, as both an expression of and response to globalization. They are also patently aware of their colonial and neo-colonial legacies, and at the same time they have deep postcolonial aspirations for themselves, for their families, for the young people of the Caribbean, for Caribbean education, for their research work and for their nations. They want to speak back contrapuntally to the metropolitan centre (Said 1993) to which they are connected historically and through flows of people, of culture, and of theory and research methodologies. Some of their research involves the prodding of 'public memory' about colonial histories and their neo-colonial present and what this means for postcolonial aspirations. They are also aware of the gap Spivak (2003) points to between their intellectual and material locations within the Caribbean and those of the less privileged, but they also see themselves in some ways distanced from the agental capacity of doctoral researchers working in the academic centres of the global north.

These doctoral students are very politically aware of the impact of US cultural industries on culture, politics, young people, and schooling in the Caribbean. They are aware of the ways in which globalization has spawned a neoliberal individualism and aspirations among the young in the Caribbean to be like rich young black Americans. These aspirations have partially replaced the independence and nationalist movements and their collectivity in the national habitus. But the students want to work with and against these matters, politically and in research terms. And for some, Cuba and Venezeula provide political alternatives. Their major aspiration for their own doctoral research is to build research capacity in the Caribbean, developing the capacity to research, rather than simply being researched, both individually and collectively. They desire the Caribbean to be more than a site for research and the application of theories and methodologies developed elsewhere. Here they talk of the ways in which donor agencies have to this point defined *who* were regarded as researchers in the Caribbean. They also see the heavy research *for* policy character of most of the research that has been conducted. They see this capacity to do research as potentially strengthening national capital in the face of globalization and international agencies. Some of the students are involved in policy making inside the state and others in the political process through teacher unions and other activist organisations. They are close to government and want educational policy research to be *for* policy as well as *of* policy and sensibly eschew this unhelpful binary. Thinking about the issues involved shows them and me how doing research can be used pedagogically to understand research and to move it towards deparochialization.

Despite all of this and the awareness that some Caribbean concepts such as 'creole' are vernacular versions of the concept of 'hybridity' central to postcolonial theory, the western, metropolitan character of citational contexts, theories, and methodologies restricts possibilities and is itself reflective of past and present geo-politics.

CONCLUSION

The argument has been advanced in this chapter that Bourdieu's theoretical work and research disposition allows for the deparochializing of educational policy research in terms of the global stretching of the research gaze and the elision of society with the national in social theory. Narrative one has shown how his work can enable such a global gaze within the field of educational policy research. Bourdieu's approach also prescribes epistemological diffidence, as called for by Appadurai, in its overt rejection of an epistemological state of perfect innocence in research and its demand for reflexivity as central to the (deparochialized) research habitus. Bourdieu's work also allows globalization to be dealt with in a non-reified way by suggesting that it is a political project reflecting past and present (colonial, neo-colonial and postcolonial) politics and resultant asymmetries of power between nations. The pedagogical anecdotes of narrative two demonstrate the desperate need for, as well as, difficulties of deparochializing educational research.

Furthermore, as I have also argued in this chapter, there have been some serious challenges to the research canon from within the western academy itself. The research edifice is perhaps not as impermeable as Appadurai's account might imply, yet there is no denying the multiple ways in which contemporary public policy about research and universities in the West at least have somewhat constrained research possibilities in terms of disciplines and topics and the intellectual resources brought to bear upon them (Ozga, Seddon and Popkewitz 2006). The challenges to the research ethic have come from several directions, including feminist, neo-Marxist, critical theory, poststructural, postmodern, narrative, life-history, and more lately postcolonial theories and methodologies (Robinson-Pant, 2005; Mutua and Swadener, 2004). The work of Tuhiwai Smith (1999) on decolonizing methodologies has been a boon to the Caribbean doctoral students, allowing a view of how things might be done and thought otherwise. In an epistemological and methodological sense, these approaches could be grouped as post-positivist and they have contributed to deparochializing educational research and challenging western metaculture.

Appadurai argues that a progressive globalization requires a destabilization of the research ethic and its paradigmatic approaches. This requires a new mix which would include ideas and perspectives from the global south, from pre-research paradigm thinkers, from student and academic flows, from NGOs, from activists, from public intellectuals, from state workers and policy makers, from humanities and social science scholars, from the post literatures and so on. In spite of these hopeful developments, the asymmetrical power realities on the world political stage will also make the globalization of knowledge and the deparochialization of research a difficult but continuing and necessary task.

NOTES

1. This chapter owes much to formal and informal discussions with my colleagues and students involved in Sheffield's Caribbean master's and doctoral programs and in the Sheffield based full-time master's programme in educational policy and research. I note here especially, Jennifer Lavia, my former colleague and Kentry Jn Pierre, my doctoral student. I have learned much from them and have had my postcolonial awareness heightened by working with them and students in the Caribbean. I also mention my debt to Wilf Carr and Paul Standish who bring a philosophical sensibility to epistemological issues of theory and research in education in the context of globalization and postcolonial aspirations. I also thank Ian Hextall, Trevor Gale, Martin Mills, Colin Symes, Kalervo Gulson, and Nicholas Lingard for their very useful comments on various drafts of this chapter and Sandra Taylor and Shaun Rawolle for helping with the thinking around Bourdieu and the emergent global educational policy space. An earlier version of this chapter was presented to the *Globalisation of the Research Imagination Conference*, Prato, Italy, 24–26 October 2005. Another version of this chapter appears in *Globalisation, Societies and Education*, Vol 4, No 2, 2006, and I would like to thank the editor for permission to publish it.
2. As an aside, we need to recognize how deeply parochial are the neoliberal political and economic agendas. Further, there is also often a deeply parochial character to US academic literature reviews where the world and the US are taken to be synonymous, not unlike the World Series in baseball.
3. For an account of vernacular globalization in educational policy production and implementation, see Lingard (2000).
4. Young (2003:1–8) sees four elements of what I would call an aspirational postcolonial politics, namely: (a) the right of all people to the same levels of material and cultural well-being; (b) contesting the reality that the nations of Africa, South America, and parts of Asia are in multiple positions of subordination to the US and Europe; (c) creating theoretical approaches which challenge dominant western ways of seeing; and (d) a reorientation towards 'the perspectives of knowledge, as well as needs, developed outside of the west' (p. 6).
5. Said (2001) in his essay, 'On Defiance and Taking Positions', sees functioning as a public memory as central to the role of the public intellectual. This is even more the case in respect of public intellectuals in postcolonial nations and is also a central element of postcolonial pedagogies.

REFERENCES

Appadurai, A. (1996) *Modernity at large*, Minneapolis: University of Minnesota Press.

—— (2001) 'Grassroots globalization and the research imagination', in A. Appadurai (ed.), *Globalization*, Durham, NC: Duke University Press.

Ball, S.J. (1994) *Education reform: a critical and poststructural approach*, Buckingham: Open University Press.

—— (1998) 'Big policies/small world: an introduction to international perspectives in education policy', *Comparative Education*, 34.2: 119–130.

Beck, U. (2000) 'The cosmopolitan perspective', *British Journal of Sociology*, 51.1: 79–105.

Bourdieu, P. (1995) 'Foreword', in Y. Dezalay and D. Sugarman (eds.), *Professional competition and professional power: lawyers, accountants and the social construction of markets*, London: Routledge.

—— (1996) *The rules of art: genesis and structure of the literary field*, Cambridge: Polity Press.

—— (1998) *Acts of resistance: against the new myths of our time*, Cambridge: Polity Press.

—— (1999a) 'The abdication of the state', in P. Bourdieu et al. (eds.), *The weight of the world: social suffering in contemporary society*, Cambridge: Polity Press.

—— (1999b) 'The social conditions of the international circulation of ideas', in R. Shusterman (ed.), *Bourdieu: a critical reader*, Oxford: Blackwell.

—— (2002) 'The politics of globalization', *Le Monde*, January.

—— (2003) *Firing back against the tyranny of the market* (Trans. Loic Wacquant), London: Verso.

Castells, M. (2000) *The rise of the network society* (2nd ed.), Oxford: Blackwell.

Dimitriadis, G. and McCarthy, C. (2001) *Reading and teaching the postcolonial from Baldwin to Basquiat*, New York: Teachers College Press.

Fisher, J. and Mosquera, G. (2004) 'Introduction', in G. Mosquera and J. Fisher (eds.), *Over here. international perspectives on art and culture,* Cambridge, MA: MIT Press.

Freire, P. (1972) *Pedagogy of the oppressed*, Harmondsworth: Penguin.

Gregory, D. (2004) *The colonial present*, Oxford: Blackwell.

Henry, M., Lingard, B., Rizvi, F. and Taylor, S. (2001) *The OECD, globalization and education policy*, Oxford: Pergamon Press.

Jayasuriya, K. (2001) 'From political to economic constitutionalism', *Constellations*, 8.4: 442–460.

Lawn. M. and Lingard, B. (2001) 'Constructing a European policy space in educational governance: the role of transnational policy actors', *European Educational Research Journal*, 1.2: 290–307.

Lingard, B. (2000) 'It is and it isn't: vernacular globalization, educational policy and restructuring', in N. Burbules and C. Torres (eds.), *Globalization and education*, New York: Routledge.

Lingard, B. and Rawolle, S. (2004) 'Mediatizing educational policy: the journalistic field, science policy and cross field effects', *Journal of Education Policy*, 19.3: 361–380.

Lingard, B., Rawolle, S. and Taylor, S. (2005) 'Globalizing policy sociology in education: working with Bourdieu', *Journal of Education Policy*, 20.6: 759–777.

Mann, M. (2000) 'Has globalization ended the rise and rise of the nation state?' in D. Held and A. McGrew (eds.), *The global transformation reader*, Cambridge, UK: Polity Press.

Marginson, S. (2004) 'National and global competition in higher education', *Australian Educational Researcher*, 31.1:1–28.

Massey, D. (1994) *Space, place and gender*, Cambridge: Polity Press.

Mufti, A.R. (2005) 'Global comparativism', in H. Bhabha and W. J. T. Mitchell (eds.), *Edward Said continuing the conversation*, Chicago: University of Chicago Press.

Mutua, K. and Swadener, B. B. (eds.) (2004) *Decolonising research in cross-cultural contexts*, New York: State University of New York Press.

Ozga, J. and Lingard, B. (2006) 'Globalization, education policy and politics', in B. Lingard and J. Ozga (eds.), *The RoutledgeFalmer reader in educational policy and politics*, London: RoutledgeFalmer.

Ozga, J., Seddon, T. and Popkewitz, T. (eds.) (2006) *Education research and policy: steering the knowledge-based economy*, London: Routledge.
Radhakrishnan, R. (2000) 'Postmodernism and the rest of the world', in F. Afzal-Khan and K. Seshadri-Crooks (eds.), *The pre-occupation of postcolonial studies*, Durham, NC: Duke University Press.
Rizvi, F. (2004) 'Debating globalization and education after September 11', *Comparative Education*, 40.2: 151–171.
—— (forthcoming) 'Postcolonial perspectives on globalization and education', *Cultural Studies, Critical Methodologies.*
Robinson-Pant, A. (2005) *Cross cultural perspectives in educational research*, Milton Keynes: Open University Press.
Said, E. (1993) *Culture and imperialism*, London: Chatto & Windus.
—— (2001) *Reflections on exile and other literary and cultural essays*, London: Granta Books.
—— (2003) *Orientalism*, London: Penguin.
Schriewer, J. and Martinez, C. (2004) 'Constructions of internationality in education', in G. Steiner-Khamsi (ed.), *The global politics of educational borrowing and lending*, New York: Teachers' College Press.
Shohat, E. (2000) 'Notes on the 'postcolonial', in F. Afzal-Khan and K. Seshadri-Crooks (eds.), *The pre-occupation of postcolonial studies*, Durham, NC: Duke University Press.
Smith, L. Tuhiwai (1999) *Decolonizing methodologies research and indigenous peoples*, London: Zed Books.
Spivak, G. C. (2003) 'Righting wrongs', in N.Owen (ed.), *Human rights, human wrongs*, Oxford: Oxford University Press.
Taylor, S., Rizvi, F., Lingard, B. and Henry. M. (1997) *Educational policy and the politics of change*, London: Routledge.
Tikly, L. (2001) 'Globalization and education in the postcolonial world: towards a conceptual framework', *Comparative Education*, 37.2: 151–171.
Urry, J. (2000) *Sociology beyond societies: mobilities for the twenty-first century*, London: Routledge.
—— (2002) *Global complexity*, Cambridge, UK: Polity Press.
Viswanathan, G. (ed.) (2001) *Power, politics, and culture interviews with Edward W. Said*, New York: Pantheon Books.
Young, R. (2003) *Postcolonialism: a very short introduction*, Oxford: Oxford Univeristy Press.

14 Trade unions, strategic pedagogy and new spaces of engagement

Counter-knowledge economy insights from Columbia

Mario Novelli

On Christmas Day in 2001 several hundred members of a public service union located in the southwestern city of Cali, Colombia occupied their company's 17th floor central administration building, the CAM Tower. The occupation was in response to the Colombian government's decision to move ahead with plans to privatize EMCALI[1] the state owned utilities company which provides electricity, water, and telecommunications to the city's two million inhabitants. For SINTRAEMCALI,[2] the public service union involved, this was yet one more challenge in their long running battle to keep their company in public hands.

During the course of the 36-day occupation, the CAM tower became the geographical headquarters of a citywide series of mass mobilizations. Tens of thousands of people in Cali went to meetings, joined marches, blocked off roads, and engaged in political protest in the defence of EMCALI as a state-owned public utilities provider. The vast majority of these were drawn from the poorest neighbourhoods in the city. This local activity was complemented by solidarity actions both in Bogotá, the capital, and in London, UK. In Bogotá, SINTRAEMCALI workers and supporters occupied the headquarters of the Superintendent of Public Services[3] for 14 hours while local public service trade unionists formed a protective cordon around the building. In the UK pickets outside the Colombian embassy were organized on several occasions in solidarity with SINTRAEMCALI workers, and two live video linkups were made between leaders of the British Trade Union Congress (TUC) and workers inside the Cali occupation. During key moments of the dispute interventions were made either via letters to the Colombian government or face to face meetings with Colombian diplomats by the British TUC, Amnesty International, Public Services International (PSI), the International Confederation of Free Trade Unions (ICFTU), UNISON, the FBU, ASLEF, War on Want, the UK based Colombia Solidarity Campaign, and a range of other UK trade union and social organizations. Messages of support and solidarity for the occupation were received by e-mail from as far afield as Papua New Guinea, Indonesia, and the Philippines.

On day 36 of the occupation the workers emerged from the CAM tower with a signed agreement from the Colombian government to freeze the privatization plans. Victory, even if only of one battle, had been achieved in an ongoing struggle that has cost the lives of 17 SINTRAEMCALI members, murdered by right wing paramilitary forces linked to the Colombian state. Many more activists have also been displaced from their homes and families, after being targeted by state and para-state forces for their involvement in the anti-privatization movement.

The question of how this victory was achieved, and that knowledge disseminated, remains important, for it was done in the face of opposition from local elites, the national government, the World Bank, the International Monetary Fund (IMF), and several key multinational corporations (MNCs) with all the structural, political, and military power that this entails. It is important beyond Colombia because similar networks of actors have managed over recent decades to implement other processes of privatization across low-income countries, often in the face of widespread opposition, and generally to the detriment of the poorest sections of these populations. And it is also important because it challenges the commonly held thesis that in the era of neoliberal economic globalization capital's new found mobility allows it an inevitable strategic advantage over labour movements due to their alleged inability to operate on the global scale (Castells 1997).

TOWARDS A COUNTER-KNOWLEDGE ECONOMY

Within global policy circles such as the IMF, the OECD (2001), and the World Bank (2003), there is increasing emphasis on 'knowledge', the 'knowledge economy', and 'policy learning', aimed at making capitalism more efficient, competitive, profitable, and dynamic. Oppositional forces, trade unions, and social movements often find themselves organizing against these objectives substituting for competitiveness and profitability goals such as social justice, equity, and equality. In this process of resistance and the development of alternatives, these movements are also engaged in knowledge production—counter-hegemonic knowledge—which plays a key role in determining success and failure. Yet research on the production, dissemination, and exchange of counter-hegemonic knowledge remains marginalized.

Recently, there have been calls to develop a 'social movement university' out of the networks of actors engaged in the World Social Forum process, which aims to draw together critical academics and social movement activists from across the world to engage in these types of counter-hegemonic research praxis, knowledge production, and dissemination (Santos 2003). This chapter seeks to contribute to those types of processes and to advocate a broadening of the research agenda on the 'knowledge economy' beyond both celebration and critique of hegemonic knowledge (Robertson 2006) towards an appreciation of the existence of alternative knowledges, alternative economies, and alternative ways of knowing and learning that offer insights into possibilities of constructing different futures from our neoliberal present.

This research agenda is part of what Santos calls 'a sociology of absences' to uncover processes that are hidden by hegemonic discourses. One of the central 'absences' within contemporary academic research is that of the 'south'—which Santos uses metaphorically as the site where those suffering under global capitalism exist and resist (Santos 1995: 506). He advocates a learning process whereby academics begin to try to 'understand the struggles that are going on around the world' through a process of 'critical engagement' with those denied the benefits of neoliberal globalization. Crucial in Santos's (2005: 3) perspective is that 'the understanding of the world exceeds considerably the western understanding of the world', yet the western domination of academia has meant that this partial understanding has come to represent the fulcrum of intellectual debate. Decentring this process and seeking out alternative knowledges thus becomes a key task in reinvigorating critical theory. This research on SINTRAEMCALI represents a modest contribution to that broader agenda.

OUTLINE OF THE CHAPTER

First it is necessary to both provide some context to the research site, and also clear the ground in terms of my own understanding of strategic pedagogy, how it operates, and how we can break away from a narrow understanding of education that limits our observations to formal classroom based learning; this will be done in the next sections. Developing from that, I will explore the concepts of space and scale and draw on Cox's (1998) notion of 'spaces of dependence' and 'spaces of engagement' as an analytical tool to highlight the linked network of associations that were constructed by SINTRAEMCALI and within which this strategic pedagogy emerged. This will then be followed by an exploration of the development of SINTRAEMCALI from the mid-1990s onward and the strategies and pedagogical 'spaces of engagement' that emerged. Finally, I will return to the CAM tower occupation to highlight how the different 'spaces of engagement' came together to produce more than the sum of their parts and ultimately proved too much for a Colombian government reeling under this relentless multi-scalar resistance.

BACKGROUND TO RESEARCH

The research is firmly rooted in the context of neoliberal structural reforms that began in Colombia in the early 1990s. The shift to neoliberalism in

Colombia led to wide ranging transformations of social relations as the economy 'opened up' to global markets. While Colombia has always been a highly unequal society, the neoliberal era intensified these inequalities. During successive governments in the 1990s a range of policies based on a classic neoliberal recipe of decentralization, liberalization, privatization, and fiscal austerity were attempted, with varying degrees of success, as workers, campesinos, indigenous communities, black communities, social organizations, students, and trade unionists contested policy and engaged in defensive struggles to hold onto the limited gains won under the previous economic model (Ahumada 1998, 2001; Sarmiento 2001). For that resistance many social movements paid a heavy price for in Colombia state/social movement relations are permeated by violence.

According to government homicide statistics Colombia has had, during the 1990s, around 25,000 violent deaths per annum (World Bank 1999: 4). While the vast majority of these deaths can be attributed to 'social' violence, according to Colombian Human Rights NGOs there are between 5,000 and 8,000 'political' murders per annum. While many of these deaths are linked to the decades-long civil war being fought out between Marxist-inspired guerrilla movements and Colombian state and "para-state" forces, the casualties go well beyond armed combatants and extend to broad sections of civil society and social movements. Within the context of what both the Colombian government and its US ally term 'a counterinsurgency war', the civilian population and non-armed civil society groups are often targeted, particularly by the far right paramilitary organizations ('death squads') linked to the Colombian state (Human Rights Watch 1996, 2000, 2001). Social movement activity and the defence of economic, social, and cultural rights is interpreted as merely a cover for guerrilla activity, and thus they are constructed as representing legitimate military targets (Stokes 2004). This 'dirty war' has led to over 2.9 million people becoming internally displaced out of a population of 45 million. Within the Colombian trade union movement, since 1986, there have been over 3,800 union leaders and activists murdered (CUT 2004) by paramilitary organizations, which makes Colombia the most dangerous place in the world to be a trade unionist.

In this context of conflict and violence researching social movements raises particular ethical and political questions. My own response was to deploy a committed research approach which combined extended ethnographic inquiry, drawing on Burawoy's Extended Case Method (1998), with active participation in the day to day work of the trade union's national and international solidarity work (Mathers and Novelli 2006; Novelli 2006b). Over a four-year period I spent 14 months working for SINTRAEMCALI in Colombia. Furthermore, in the UK I worked with colleagues in the Colombia Solidarity Campaign to build links between British trade unions and social movements. This approach built up considerable mutual trust and allowed me open access to SINTRAEMCALI's leaders and activists and the union's archives. It also provided me with the opportunity to play my own small part in a social struggle that I became committed to. By moving back and forth between Colombia and the UK the research approach also facilitated a multi-sight perspective on the possibilities for new 'labour geographies' (Herod 2001) and rich empirical evidence of how labour movements were learning new tactics, acting 'beyond borders' (Keck and Sikkink 1998) and developing new strategies to challenge the neoliberal restructuring taking place throughout the world.

STRATEGIC PEDAGOGY AND LABOUR MOVEMENT LEARNING

Despite an important and growing literature on trade union renewal and resistance to neoliberalism (Herod 1998, 2001; Lambert and Webster 2001; Moody 1997; Munck 2002; von Holdt 2002; Waterman 1998), little intellectual inquiry has been devoted to exploring the micro processes of how trade unions transform their activity and the role played by the processes of education and learning. Part of the reason for this absence lies in a narrow conceptualization of education as taking place only in classrooms or through formal means. My argument, however, is that trade unionists, as political activists, learn a great deal through their activities, organizational and political work, and that this praxis is an educational process in itself which sits alongside more 'formal' learning spaces. Processes of 'education' and 'learning' are thus intimately tied to broader social processes, and can only be understood within that context. The question then is how best to understand the conditions under which trade unions and their members might change focus and direction and shift towards new patterns of organization and solidarity practices.

Pedagogy and the production of conformist or rebellious subjectivities

Pedagogy is usually considered as a lower order concept referring to sets of practices of upbringing and education, methods and processes of transmission and acquisition that seek to develop knowledges, skills, and moral order. However, as Bonal and Rambla (2003: 172) note drawing on the intellectual insights of Basil Bernstein, 'each pedagogy could and should be studied by looking at the social forces that induced, maintained and legitimated it'. Dominant pedagogies are thus studied with the aim of understanding how communication systems might produce individual and collective subjectivities, as well as ways of understanding and interpreting the world.

In developing the concept of 'strategic pedagogy' I intend to explicitly link pedagogy to the particular social forces that both produce and defend these practices. Hence pedagogies within state education systems could be seen as being related to particular modes of production and divisions of labour within societies. The pedagogic device, the condition for the materializing of symbolic control, is the object of a struggle for domination, for the group who appropriates these devices can influence consciousness, identity and desire. The question is pedagogy in whose interests, or for what consciousness, desire, and identity (Bernstein 1999: 269)?

This leads us to the task of inquiring into the different forms of 'socialization, education and work that promote rebellious or, on the contrary, conformist subjectivities' (Santos 1999: 41). Focusing on contemporary conditions of neoliberal globalization, and drawing on Bernstein's concept of the Totally Pedagogized Society (TPS), Bonal and Rambla (2003: 174, citing Bernstein 1996: 72) go on to explore how 'fast capitalism' necessitates the worker being continually in a process of reforming and reskilling to 'cope with the new requirements of work and life'. Gone are the securities of stable occupations. A pedagogy of 'trainability' is therefore deployed to propagate and justify the continual need for reskilling. This form of neoliberal subjectivity transfers responsibility from the state to the individual by pressing the individual to continually upgrade their skills and adapt to the shifting needs of the market. While this is an important aspect of pedagogy, the process extends to other spheres of social life, conceptualizations of the individual, of rights, and of values, of consumption habits, and of instilling the 'commonsense' conceptualization that there is no alternative to the contemporary status quo.

While these new pedagogical modalities serve to produce conforming subjectivities, what might an alternative pedagogy producing rebellious subjectivities look like? And what would be its social base? Within this understanding of pedagogy related to particular social forces with particular goals and objectives we can begin to explore what the strategic pedagogy of SINTRAEMCALI looks like, the nature of the social forces that give it life, and what particular pedagogical modalities were deployed in the construction of oppositional strategies. In order to begin this we need to look for their roots in the 'popular education' movement.

Popular education in Latin America as knowledge resource

Latin America provides a fertile ground for the search for models of education that can be deployed to generate critical consciousness. 'Popular Education', initially proposed by Paolo Freire, has captured the imagination of many social movements seeking to challenge elite structures of domination. In Nicaragua, the work of 'popular educators' prepared the ground for the 1979 Sandinista revolution (Arnove 1986), and one can see the legacy of Freire's work being continued throughout the continent.

Popular education is thus seen as one of the vehicles through which the process of challenging unequal structures can be achieved (Kane 2001).

It has, at its centre, a fundamental commitment to social change in the interests of the 'popular' classes. Furthermore, there is a direct relationship between this type of education and the institutions and organizations that have historically emerged to defend the interests of the poor, such as trade unions and social movements, and it seeks explicitly to strengthen these movements (Jara 1989 cited in Kane 2001: 9). This organic relationship means that the 'organization' becomes the 'school' in which popular education takes place, and their 'struggles and actions, their forms of organisation, their "culture," in the broadest sense, constitute the starting point of popular education and its on-going field of enquiry' (Kane 2001: 13).

In that sense, 'popular education' needs to be seen as not only involving formal educational events but as part of much bigger processes which, though appearing 'informal' and 'arbitrary', are very deliberate. In this definition, both the 'popular education' events that take place, and the actual practice of 'strategy development' and 'protest actions' can be seen as examples of popular education, whereby the 'school' (the social movement) learns. The first occur when people consciously engage in educational practices (schooling), and the second when people are learning through social action. Foley (1999) suggests that a broad conception of education and learning should include formal education (taking place in educational institutions), incidental learning (taking place as we live, work and engage in social action), informal education (where people teach and learn from each other in workplaces, families, communities, social movements), and non-formal education (structured systematic teaching and learning in a range of social settings).

If we are to explore these processes, then we need to extend our gaze beyond formal training courses for activists and develop an analytical framework that is 'open' and which allows for the rich diversity of ways in which social movements (their activists and supporters) engage in learning.

Strategic pedagogy as knowledge production in action

In order to do this we need to make operational concepts of learning and education that take place within time and space through struggle and contestation. What is needed is a dynamic model that can capture the dialectics of this process. Hay (1995) talks of 'strategic learning' and offers a useful diagram to highlight this as a process in motion (Figure 14.1).

Hay's model provides a way of theorizing the process whereby a trade union engages in strategic analysis, which in turn leads to strategic action, then to intended and unintended consequences of action, and to further reflection/analysis and action—a strategic pedagogy. The agency of the trade union itself is immersed in structural constraints, which provide possibilities as well as limitations (Cox 1996), while the strategic action often changes the structures upon which it acts. Throughout these entire processes, incidental, formal, informal, and non-formal education will be taking place

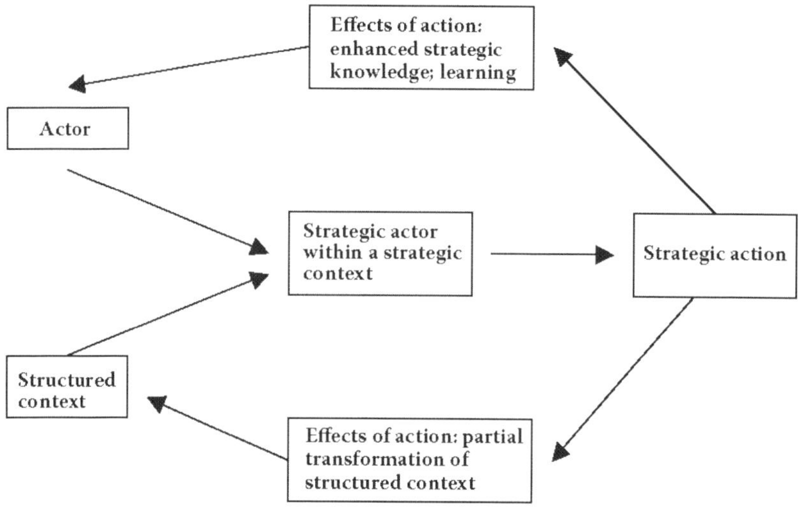

Figure 14.1 Hay's (1995) model of 'strategic learning'.

and influencing strategy and outcomes to differing degrees. Hay's model provides a way of exploring how a trade union strategically develops.

SPACES OF DEPENDENCE AND ENGAGEMENT

Having laid out an understanding of 'strategic pedagogy' I now want to turn to questions of scalar politics and the spatiality of resistance. In order to understand why space and scale matter it is necessary to step back and consider the way neoliberal reforms over the last three decades have restructured the geography of global capitalism. Much of the literature on economic globalization notes how 'capital' in the wake of the economic crisis of the 1970s sought to break free from the nation-state regulation that it was previously bound to (taxation, national subsidies, import controls and tariffs) (Brenner 2004). This shift beyond national borders has meant that capital has sought new forms of regulation to manage the expanding accumulation process. As the prior stage of Fordist accumulation was based on national economies and inter-state relations, the new emerging form of regulation needed to find mechanisms beyond the nation-state. This shift entailed a range of processes of rescaling and shifting the spatial relations of political, social, and economic accumulation. The shifting scales of accumulation, together with advances in technology, have redrawn the parameters for governance, power, and control and have important implications for strategies of contestation and resistance covered in this chapter.

One of the crucial elements of capitalism relates to the changing relationship between space and time, 'time-space compression', and 'fixity and motion'. These involve conflicts between the need for capital to circulate globally, and also to be located in fixed territorial spaces (Brenner 1999; Harvey 1989) that require management in order for accumulation to take place. Conflict can arise over the means of regulating these new spaces. Scale is produced through this process of negotiation, between the need to move to different geographical locations and the need to be fixed in a particular place to produce goods. Neoliberal globalization has broken down old 'spatial fixes', 'settlements', or 'social contracts' and sought to forge new ones. We can interpret the conflicts over attempts to 'fix' new territorial pacts such as the EU or the Free Trade Area of the Americas (FTAA) as battles between different social forces that may benefit or lose out in new territorial arrangements. Likewise we can see the shift to both the local scale with increased policies of 'decentralization' occurring simultaneously with a rise in the importance of supranational organizations such as the IMF, World Bank, and the WTO. At these different scales we can see competition between different countries over the type of future Europe or Latin America under construction, between corporations, between different sections of capital, and between labour.

Unlike a conventional understanding of scale (e.g. Hart 1982) which perceives scales as being fixed or pre-given (i.e. the local, national, and regional as permanent entities), research in economic and political geography (Brenner 1999; Cox 1998) has conceptualized 'scale' as being actively constructed for particular purposes by different social forces. Scales are thus arenas for social struggle that are fought over, expanded, and constructed. Hence, shifts in scale both upwards and downwards are means to shift contradictions and the potential for conflict to different locations where different social forces might be privileged. As Harvey (1989) notes, there are differential relations between different classes at different scales, and some social forces are better able to command and control certain spaces. This leads us to Castells's (1997) argument that because of lack of mobility labour is, and labour movements are, likely to fade away as capital in hyper mobile and fluid forms shift around the globe.

Yet Castells's understanding is a highly static and undialectical one and presupposes the inability of labour to organize beyond local levels. As Herod (2001) has noted, the power of capital and labour is not static—the power of both capital and labour may change over time. Furthermore, the notion that capital is global and labour fixed fails to recognize either the varied terrain of the global—with different relational forces present in different places—or that the global is also localized in some particular place where its power relations may be differentialized (think of global production chains with each component locally based). Ultimately, while it may be true that capital's new found mobility has strengthened its structural power, it still has susceptibilities and labour has the potential to alter its own resources through strategic change.

While any particular era places limits on the possibilities for change it also provides strategic possibilities, and it is these strategic possibilities that are picked up by Kevin Cox (1998) in a paper that begins to unpack how a politics of scale can serve as a mode of resistance. In Cox's (1998) paper on 'Spaces of Dependence and Spaces of Engagement' he discusses this politics of scale by making an analytical distinction between 'spaces of dependence'—spaces that are to be defended and strengthened, and 'spaces of engagement'—which are sites that different social actors draw upon to defend a 'space of dependence'.

Spaces of dependence are defined by those more-or-less localized social relations upon which we depend for the realization of essential interests and for which there are no substitutes elsewhere; they define place-specific conditions for our material well-being and our sense of significance. These spaces are inserted in broader sets of relationships of a more global character that constantly threaten to undermine or dissolve the smaller spaces of dependence. People, firms, state agencies, etc., organize in order to secure the conditions for the continued existence of their spaces of dependence, but in so doing they have to engage with other centres of social power, local government, the national press, perhaps the international press, for example. In so doing they construct a different form of space, which I call here a space of engagement: the space in which the politics of securing a space of dependence unfolds. This may be on a more global scale than the space of dependence, as per the idea of 'jumping scales' but it is not necessarily so (Cox 1998: 2).

What I will explore in the remainder of this chapter is how SINTRAEMCALI, through an awareness of the scalar possibilities of resistance, was able to 'jump scales' and in doing so highlight the possibilities for multi-scalar labour resistance. SINTRAEMCALI constructed a network of alliances across a range of geographical scales, which were aimed at securing, defending, and extending particular local 'rights'. The realization of these local 'rights'—to work in EMCALI, to deliver and consume public services, to adequate human rights and labour protection—were tied to a particular physical locale within which the members of the movement existed and resisted, and it is this 'space' that can be understood as a 'space of dependence'. The network of alliances that were constructed to defend this 'space of dependence' can be understood as 'spaces of engagement', which could potentially be on the same scale but could also link a range of actors and institutions on a range of scales from the local to the global. In SINTRAEMCALI's case these 'spaces of engagement' were complex and multi-scalar. Extrapolating from Cox's (1998) categorization I want to argue that these 'spaces of engagement', these 'networks of alliances', were also pedagogical spaces that required the development of particular skills and the forging of particular relationships in order to maximise the potentiality of success along each node of the network—this represented the content of the strategic pedagogy.

SINTRAEMCALI'S STRATEGIC PEDAGOGY AND ITS SPACES OF ENGAGEMENT

The push to privatize EMCALI began in the mid-1990s. Prior to that, according to Alexander Lopez, the ex-President of the trade union and architect of its renewal, SINTRAEMCALI 'was just a calm, normal union working on normal grievances' (Podur 2003). However, the onset of neo-liberal restructuring in Colombia produced a new environment that forced the organization to rethink its strategic direction. In response a new and more militant group of workers began to take the initiative. The new leadership emerged from a rank and file committee that was born in the sewerage section of the company. After initially gaining success in improving working conditions for the sewerage workers, the rank and file committee began to push for greater influence in the trade union as a whole. In 1994, Alexander Lopez was elected to the Executive Committee (EC) and within three years the balance of power within the trade union was won over to a position of radical opposition to privatization. From that point on the question became one of strategy. How could the union take on a highly repressive state and its transnational allies in the complex and violent conditions of Colombia? Four linked strategic directions emerged which relied on the construction and strengthening of particular spaces of engagement.

Worker/community alliance

One of the crucial strategies that emerged was the necessity of reframing the conflict over public services away from a trade union/state problem to one between the Cali community and the state. This necessitated a broad strategic alliance with local poor communities, but constructing this was not an easy task. There was a problem within EMCALI itself that could not simply be explained by corrupt management practices. A certain level of complicity existed between workers and managers concerning the weak performance of the company. This poor 'service ethic' served to legitimate government and media accusations that the workers benefited from the good salary and benefits provided by the company at the expense of the community. There was a need to change the attitude of workers, and the attitude of the local community towards them. This required a sustained process of education on both sides, and a meeting point whereby workers and local communities could join together (Interview with Alexander Lopez 2002).

This process of building up relations with the local community began in the mid-1990s, and was aimed at raising the 'political consciousness' of the local marginalized communities and instilling EMCALI workers with a sense of civic pride and responsibility. This work ranged from organizing public meetings, to workshops, to joining local community demonstrations and protests, and to setting up a formal technical school based at the union's headquarters, aimed at teenagers unable to afford secondary education. One central initiative was setting up 'Minga Comunitarios' (an indigenous phrase meaning 'to come together'), whereby all EMCALI workers would give up one weekend per month to carry out repairs and damage to infrastructure in the poorest areas of the city. In order to cement this trade union/community alliance new forms of democratic representation emerged whereby the trade union and community activists could develop joint proposals on strategy for the defence of public services. All of these initiatives served to develop strong bonds between workers and local communities.

Alexander Lopez talks of the obstacles that the trade union initially faced in these pedagogical processes, the objectives of the pedagogical interventions and the attitudinal transformation that later prevailed:

> In the beginning we had resistance from the community, because the community saw the workers as the problem, and we explained to the community who are the real problem, and where the neoliberal politics comes from, what is the World Bank, what is the International Monetary Fund, cause the people don't know this. It has been a very didactic strategy, very long, but also very beautiful and fruitful because...the community takes time to make a decision, but when it makes it, it makes it forever, I believe that the people, yes, they have this logic, they don't recognize a lot of things, but they feel it when you speak, when you explain, and these type of lovers are not temporary loves but lovers that last forever. (Interview with Alexander Lopez 2002: 14)

The visible presence of the trade union in the neighbourhoods served the purpose of 'conscientization' in Freirean terms (Freire 2000), where a 'dialogical' process begins to emerge between organized labour and communities. As Alexander Lopez notes (Interview 2002: 14), apart from attending to the real material needs of local communities, and raising political consciousness, these were aimed at providing a 'space' for challenging the attitudes of the communities and the workers towards each other.

Returning to the concrete outcomes of this 'space of engagement' we can see that in the process of the development of the trade union/community alliance the community brought with it a range of material and ideational resources that assisted the SINTRAEMCALI struggle. This network of organizations and individuals to differing degrees would show solidarity with the union's struggle, engage in marches, public demonstrations, provide material and political support to the union, apply pressure on local political representatives, and disseminate further an alternative reading of particular disputes. On many occasions they blocked roads, joined marches, attended public meetings, and showed solidarity with the union in a range of ways, not least in defending the perimeter of the series of occupations, and particularly the 1998 and 2001 occupations.

'PLAN PARE'—ALTERNATIVE DEVELOPMENT STRATEGY

From the early 1990s SINTRAEMCALI developed a collective understanding of the defence of public services as a question of the defence of national sovereignty. However, what was lacking was a negative critique of neoliberal arguments on the inefficiency of state owned public services, and a positive critique of how public services within EMCALI could be made more efficient. This led to the development of the 'Plan PARE'[4] for the recuperation of the company—a blueprint for the efficient management and delivery of public utilities that was developed by a combination of SINTRAEMCALI workers, sympathetic local technicians, lawyers, and financial experts. The plan had three phases: The first period between 1995 and 1999 focused on investigating and exposing corruption within the company and highlighting the fact that the crisis the company was facing was due to widespread management corruption, political interference, and incompetence. The second stage, from 1999 to 2001 resulted in the worker led development of an alternative management plan for the company. This was presented in late 1999 by SINTRAEMCALI but not accepted either by EMCALI management, the local council, or the Superintendent of Public Services. However, this was to change in 2001 when local elections led to a change in administration and a commitment by the new Mayor to keep EMCALI in the public domain and to implement the plan. The third stage of the development of Plan PARE was its implementation beginning in May 2001. Within eight months EMCALI had begun to show impressive results in cutting costs, improving efficiency, and reducing the company's debt. While SINTRAEMCALI celebrated the reversal of economic fortune there was deep anxiety in government circles about the example EMCALI might be setting. On 24 December 2001 the national government sacked the new management team who had implemented the Plan PARE and announced intentions to push through with EMCALI's privatisation (Interview with Nelson Sanchez 2002: 9). SINTRAEMCALI's response was the occupation that began the following day.

Addressing the issue of the management of public services through a critique of corruption and the pro-active Plan PARE allowed the union to enter into a whole series of new 'spaces of engagement'. In this new space the question of 'who manages, who controls' was once again placed on the negotiating table (Interview with Alexander Lopez: 22). SINTRAEMCALI began to reconceptualize its role away from one of mediation to one of worker control of production. This shift from mediation to control, while remaining partial, was a highly emancipatory shift as it began to generate questions not only of workers' rights and benefits, but also of workers' control of production and the challenging of the labour/capital relation.

The development of the Plan PARE also represented an intense laboratory whereby workers developed the new skills and knowledge necessary to manage the company and develop alternatives. Crucial to this process was the culture of learning that emerged within SINTRAEMCALI, and particularly the developing interest of workers in learning every intricate detail about the functioning of the company. The Plan PARE action research project meant that the union had, by 2001, developed an enormous knowledge of the day to day running of the company and its financial difficulties and was able to take its well-researched alternatives to a wide range of fora and debate alternatives.

During these interventions, networking and contacts were developed through which joint political projects could be discussed and strengthened with independent political forces. Several sympathetic representatives of Congress and the Senate supported the case against privatization. Furthermore, at the Municipal level support was given by the union to the mayoral campaign of Jon Maro Rodriquez, who upon election in 2000 began to implement the Plan PARE. The mayor was able to do this because SINTRAEMCALI was able to deliver a constructive alternative development plan for the management of the company, and a work force that was prepared to make a great many sacrifices to implement the plan successfully.

Mobilization strategy

The third strategic trajectory of SINTRAEMCALI recognized the conflictual nature of any alternative development plan and the need to apply pressure on the government. The union recognized that the social forces intent on privatizing EMCALI would not be persuaded by words alone, and thus mass mobilization and direct action had to be deployed. SINTRAEMCALI's mobilization tactics included a combination of mass based demonstrations, road blockades, and mass meetings that often brought sections of the city to a standstill. However, the trade mark of the union from 1994 became the 'occupation' of high profile buildings, a highly symbolic tool which focused attention on an issue, raised the question of 'control', but did not inflict material damage on the company, nor cause unnecessary inconvenience to the general public. The most spectacular of these were the occupations of the CAM Tower in 1998 for 16 days and in 2001 for 36 days. Both actions resulted in overturning decisions to privatize the company.

While the success of the mobilizations relied on local community support they also depended on a hard core of members and supporters prepared to put themselves at great personal risk. As noted earlier the internal transformation of the union was driven and sustained by a group of several hundred workers who emerged in and around the 'Comite de Base' (rank and file committee). These members had a strong ideological commitment to the union's objectives and to a broader process of political change within the city and country. This group were largely affiliates of SINTRAEMCALI but also included several local trade unionists from other companies, community leaders, leftist, and radical human rights activists who saw the defence of public services as a key domain of struggle. From this group come the 17 assassinated members, the forcibly displaced, and those who have been personally affected by the policies of state and para-state repression. Many were actively involved in the construction of the Plan PARE, and would have been involved in the militant actions such as the occupations, blockades, strikes etc., that developed after 1994. They would have developed great experience in organizing an occupation, their responsibilities, duties, the psychological discipline necessary, and so forth.

When asked why this group were prepared to risk their lives for the defence of public services and thus for the union, Luis Hernandez, President of SINTRAEMCALI responded:

> I believe that you cannot buy this kind of loyalty, but you must construct it, and you construct it through a strong and responsible class leadership. I believe that those of us who have been leading this organisation have demonstrated this and have built up a consciousness among the workers. This is a long process, but effective, very effective. Those workers that have accompanied us in the different direct actions have felt the strength of our organisation which is expressed in the beautiful phrase 'hasta la vida misma'. We have been prepared to do this for a collective objective, as is the defence of public property, and defending the interests of the people. (Interview with Luis Hernandez 2002: 8)

He highlights the role of the range of educational processes that many of these activists underwent:

> This loyalty has been developed from the training that our followers have received, and will continue to receive, because the good leader is in a evolutionary process, we are all in an evolutionary process and we are all learning, and developing and we will never have the final truth revealed, every day we need to learn more. That is why we are convinced that we need to generate more training and education. So that tomorrow it is not 500 workers but we can double that number or triple it, in the sense that we can keep generating more consciousness to struggle amongst the workers. (Interview with Luis Hernandez 2002: 9)

Here if we reconceptualize the processes of struggle as processes of education we can then see the importance of this in the formation of the identities of these workers. Many had had friends assassinated since 1997, and had forged strong bonds with those comrades with whom they had engaged in direct actions and occupations. During these occupations, intense solidarities and bonds of trust were forged. As one worker noted 'in the occupations we shared everything, if we had an empanada (small pie) we would divide it in six or seven pieces, we shared our thoughts, our fears and our lives' (Interview with Asdrubal Gonzalo 2002: 9). This sense of collectivity and commitment provided the unpredictable element for the Colombian state because these people had lost their fear, or at least had learned to manage it. For our understanding of the critical pedagogy of SINTRAEMCALI it was this bonding and commitment along with the new knowledge that developed during occupations that gave SINTRAEMCALI its vital edge in controlling those new spaces of protest.

SINTRAEMCALI AND HUMAN RIGHTS AND SOLIDARITY NETWORKS

The union's determination to mobilize and occupy to defend EMCALI and its uncovering of high-level corruption placed it in direct confrontation with sections of the political and economic elite. In Colombia this meant facing the full wrath of state and para-state repression. The 17 members assassinated, the false imprisonment of workers under charges of terrorism, and a general and persistent barrage of death threats and assassination attempts led the trade union to adopt a range of preventative measures and work closely with national and international human rights organizations and networks.

In 1998 a prominent human rights activist, Berenice Celeyta, set up a human rights department inside the union to monitor human rights violations and to develop a strategy of human rights defence. At the local level, the human rights department trains local activists in human rights norms and mechanisms for self-defence. The training provides participants with methods of intervention into conflicts, how to set up a local human rights network among social movements, and a background into the human rights legal frameworks that exist at regional and global levels. These human rights training courses were expanded across the region in 2002 after securing international funding for the project. These graduates in human rights courses and diplomas are able to act as intermediaries with state forces, government representatives, and international bodies in specific conflicts, and are aware of the functioning of this global infrastructure.

At the national and international level the human rights department has facilitated the involvement of major human rights organizations in SINTRAEMCALI's situation, and representatives of Amnesty International and Human Rights Watch have intervened several times. The involvement of Amnesty and Human Rights Watch is understood by the union's leaders as providing a cordon of protection within which they carry out their activities and a means of addressing and revealing Colombia's 'parallel state' of covert repression (HRW 1996, 2000, 2001; Interview with Berenice Celeyta 2002).

Closely related to this network is the role of the supranational labour and human rights bodies, such as the International Labour Organisation (ILO) and the Inter-American Human Rights Court (IAHRC). SINTRAEMCALI, with the assistance of national human rights organizations, took their case of systematic persecution from state and para-state forces to the Inter-American Human Rights Court. On 21 June 2000, the court found the Colombian government responsible for failing to provide adequate security for the trade union to carry out its internationally legitimate duties. The court ordered the Colombian government to provide protection for the entire leadership of the union including bodyguards, weapons, bulletproof vehicles, and communications equipment (IAHRC 2000). This protection, while not eliminating the danger, at least allowed some 'space' for activity and was a severe political defeat for the Colombian government, which vigorously opposed the application and denied any responsibility. The ability, experience, and success of Colombian human rights organizations to prepare cases for these processes are a great source of concern for the Colombian government.

More traditional international trade union activity was also developed alongside the Colombian CUT (the biggest of the national trade union federations), and other national trade union organizations. With the backing of the British TUC, it has also put forward motions for the last three years at the annual ILO conference in Geneva to call for a 'Special Commission of Inquiry' into the human rights emergency for Colombian workers. The motion was defeated each time by intense lobbying by the Colombian government.

The human rights department has also been a fundamental catalyst in building transnational links with trade union, solidarity, and other activist organizations in the North. In the United Kingdom, strong international links began during Alexander Lopez's brief period of exile in London from October to December 2000. Since that point there have been delegations, exchanges, and several invitations for SINTRAEMCALI representatives to talk at trade union conferences. SINTRAEMCALI forged strong links with the Colombia Solidarity Campaign, UK, which campaigns on human rights and political issues in Colombia, and several of its members have worked as volunteers in the human rights department of the union. Since 2000, there has been an ongoing process of communication and regular delegations that have forged strong bonds between Colombian and British activists. This strategy has extended more recently to Spain, Canada, the United States, and Australia and includes links with sympathetic politicians, human rights groups, NGOs, and social movements in each of these countries. The rapidly expanding Colombian exile community in these countries has strengthened the transnational links and many are founding members and activists in solidarity organizations, as is the case in the UK.

Within the development of solidarity with SINTRAECALI we can see the range of solidarity actions operating in both the more traditional forms and models of international trade union activity and the 'new' labour internationalism, anti-capitalist and human rights domains:

> In the last three years we have had ten international delegations: Amnesty International, American Watch, who carried out important work in verifying denouncements of human rights violations, three UK delegations of trade union leaders, where organizations such as UNISON, the FBU, ASLEF, Colombia Solidarity Campaign, and War on Want have participated. Two delegations from the Spanish, Red de Hermanidad (Brotherhood Network); one Canadian delegation, another Spanish one, and one mission from Public Services International which included 12 delegates from different countries, and most recently a delegation from Austria. We have seen the strengthening of this accompaniment, beginning initially with trade unions from England and later from a range of European countries, the United States, and Canada. We have received humanitarian support and emergency funds in those moments of serious crisis from UNISON, the FBU, the TUC, and War on Want. We have also been present at the World Social Forum in Porto Alegre and at the European Social Forum in Firenze (Interview with Berenice Celeyta 2003: 6).

The human rights department has provided a permanent point of contact for all this international work. This involves coordinating activities with international organizations, producing and disseminating videos, reports, and urgent actions. The vast array of contacts and experience of Berenice Celeyta, the prominent Colombian human rights activist, has ensured that the union capitalized on the potential interest of the international community in the Colombian situation.

The human rights department provided the technical skills and abilities to enable SINTRAEMCALI to challenge the Colombian state at a range of levels from local legal processes to the Inter-American Human Rights Court. Crucially, the department continues to provide training for local trade union and community activists thus transferring these vital skills. Central to the strategic pedagogy of SINTRAEMCALI was the need to create the necessary skills, knowledge, and materials so as to be able to operate effectively in these new solidarity and human rights spaces. The human rights diplomas provided the skills for trade union and community activists, the production of videos, the detailed collection of human rights testimonies, and the support for delegations both to and from Colombia, all reflected the trade union's commitment to educating both their own members in human rights and also the international community of trade union and political activists about what was occurring in Cali and in Colombia more generally. This was a highly strategic and active process that, as we saw during the CAM Tower Occupation, produced important results.

CONCLUSION

Long and hard strategic work devoted to constructing these different 'spaces of engagement' and the pedagogical processes that gave them life was central to the success of SINTRAEMCALI. These pedagogical processes took place in a variety of ways ranging from formal human rights classes to the processes of incidental learning on demonstrations. Crucially, these pedagogical processes extended out from the SINTRAEMCALI membership to local communities, activists, and international actors. While SINTRAEMCALI leaders often articulate this in the fixed scalar levels of the local, national, and international, Cox's 'spaces of engagement' allow us a way of pinpointing agency in these scales by focusing on the particular groups and actors involved and understanding the processes through which they learned to act effectively and construct their spaces of engagement. Crucially, it was not all the international community, nor national or local, but particular sets of people that carried out certain things, and their actions were linked in a myriad of different ways.

Returning to the CAM Tower Occupation, what we can now see is how these different spaces of engagement—the trade union/community alliance, the militancy of the rank and file, the management space of the Plan PARE, and the transnational solidarity space of human rights and labour rights—operated together to produce more than the sum of their individual parts and in combination, on this occasion, proved to be too much for the Colombian state, which retreated to plan for the next confrontation.[5]

While an extreme case, the knowledge unearthed in this research hopefully sheds new light on the strategic possibilities of labour in the era of neoliberal globalization to construct a different form of politics and a new spatiality of resistance able to challenge the forward march of capital. If so, then central to that process must be a renewal of interest in the pedagogical processes of trade union learning and education that can increase the knowledge base upon which alternative strategy is constructed and new spaces of engagement commanded.

NOTES

1. EMCALI is the acronym for the Municipal Enterprises of Cali, Empresas Muncipales de Cali
2. SINTRAEMCALI is the acronym for the Trade Union of the Municipal Enterprises of Cali, Sindicato de Trabajadores de las Empresas Municipales de Cali.

3. A Colombian government ministry that is responsible for the regulation and control of public services in Colombia.
4. PARE is the acronym for EMCALI Assistance and Recuperation Plan.
5. See Novelli (2004a, 2004b) for a much deeper coverage of this whole issue, and Novelli (2006a) for a stronger political and economic analysis located in the research on social movement unionism. The research was based on a four year Economic and Social Research Council funded study which included 9 months of participatory ethnographic research in Colombia.

REFERENCES

Ahumada, C. (1998) *El modelo neoliberal*, Bogota: El Ancora Editores.

—— (2001) 'Una década en reversa', *Que esta Pasando en Colombia?*, Bogota: El Ancora Editores.

Arnove, R. F. (1986) *Education and revolution in Nicaragua*, New York, London: Praeger.

Bernstein, B. S. J. (1999) '"Pedagogy, identity and the construction of a theory of symbolic control": Basil Bernstein questioned by Joseph Solomon', *British Journal of Sociology of Education*, 20.2: 265–279.

Bonal, X. and Rambla, X. (2003) 'Captured by the totally pedagogised society: teachers and teaching in the knowledge economy', *Globalisation, Societies and Education*, 1.2: 169–184.

Brenner, N. (1999) 'Globalisation as reterritorialization: the re-scaling of urban governance in the European Union', *Urban Studies*, 36.3: 431–451.

—— (2004) *New state spaces: urban governance and the rescaling of statehood*, New York: New York University Press.

Burawoy, M. (1998) 'The extended case method', *Sociological Theory*, 16.1: 4–33.

Castells, M. (1997) *The power of identity*, Malden, MA: Oxford: Blackwell.

Cox, K. R. (1996) *Approaches to world order*, Cambridge: Cambridge University Press.

—— (1998) 'Spaces of dependence, spaces of engagement and the politics of scale, or: looking for local politics', *Political Geography*, 17.1: 1–23.

CUT (2004) 'Informe Sobre Los Derechos Humanos 2004', Bogota.

Foley, G. (1999) *Learning in social action: a contribution to understanding informal education*, London: Zed.

Freire, P. (2000) *Pedagogy of the oppressed*, New York: Continuum.

Hart, J. (1982) 'The highest form of the geographer's art', *Annals of the Association of American Geographers*, 72.1:1–29.

Harvey, D. (1989) *The condition of postmodernity: an enquiry into the origins of cultural change*, Oxford: Blackwell.

Hay, C. (1995) 'Structure and agency', in D. S. Marsh (ed.), *Theory and methods in political science*, Basingstoke: Macmillan.

Herod, A. (1998) *Organizing the landscape: geographical perspectives on labor unionism*, Minneapolis, London: University of Minnesota Press.

—— (2001) *Labor geographies: workers and landscapes of capitalism*, New York: Guilford Press.

Human Rights Watch (HRW) (Arms Project) (1996) *Colombia's killer networks: the military-paramilitary partnership and the United States*, New York: Human Rights Watch.

—— (2000) *COLOMBIA The ties that bind: Colombia and military-paramilitary links*, 12.1(B): February.

——— (2001) *The 'Sixth Division' military-paramilitary ties and U.S. policy in Colombia*, New York: Human Rights Watch.

Inter-American Commission of Human Rights (IACHR) (2000) *Third report on the human rights situation in Colombia*, Organisation of American States, OEA/Ser.L/V/II.102, Doc. 9 rev. 1, 26 February.

Interview with Alexander Lopez (2002) carried out by M. Novelli, March, 2002.

Interview with Asdrubal Gonzalo (2002) carried out by M. Novelli, March, 2002.

Interview with Berenice Celeyta (2002) carried out by M. Novelli, March, 2002.

Interview with Berenice Celeyta (2003) carried out by M. Novelli, June, 2003.

Interview with Luis Hernandez (2002) carried out by M. Novelli, March, 2002.

Interview with Nelson Sanchez (2002) carried out by M. Novelli, March, 2002.

Kane, L. (2001) *Popular education and social change in Latin America*, London: Latin America Bureau.

Keck, E. M. and Sikkink, K. (1998) *Activists beyond borders: advocacy networks in international politics*, Ithaca, NY: Cornell University Press.

Lambert, R. and Webster, E. (2001) 'Southern unionism and the new labour internationalism', *Antipode*, 33.3: 337–362.

Mathers, A. and Novelli, M. (2006) 'Researching resistance to neoliberal globalisation: engaged ethnography as solidarity and praxis', *Globalizations*, 3.4.

Moody, K. (1997) 'Towards an international social-movement unionism', *New Left Review*, 225: 52–72.

Munck, R. (2002) *Globalisation and labour: the new great transformation*, New York: Zed Books.

Novelli, M. (2004a) 'Trade unions, strategic pedagogy and globalisation: learning from the anti-privatisation struggles of SINTRAEMCALI', Unpublished Ph.D. diss, University of Bristol.

——— (2004b) 'Globalisation, new labour internationalisms and education: the role of strategic learning in the transformation of the Municipal Workers Union of EMCALI', *Globalisation, Societies and Education*, 2.2: 161–190.

——— (2006a) 'SINTRAEMCALI and social movement unionism: a case-study of trade union resistance to neoliberal globalisation in Colombia' in A. Gamble et al. (eds.), *Labour, the state, social movements and the challenge of neoliberal globalisation*, Manchester: Manchester University Press.

——— (2006b) 'Imagining research as solidarity and grassroots globalisation: a response to Appadurai', *Globalisation, Societies and Education*, 4.2: 275–286.

OECD (2001) *Second OECD conference on women entrepreneurs in SMEs : realising the benefits of globalisation and the knowledge-based economy*, OECD.

Podur, J. (2003) 'Colombia's public services: an interview with Alex Lopez of SINTRAEMCALI' by Justin Podur, *En Camino*.

Robertson, S. (2006) 'Absences and imaginings: the production of knowledge on globalisation and education', *Globalisation, Societies and Education*, 4.2: 303–318.

Santos, B. D. S. (1995) *Toward a new common sense: law, science and politics in the paradigmatic transition*, London: Routledge.

——— (1999) *On oppositional postmodernism*, London: Zed Books

Contributors

Felicity Armstrong is a senior lecturer at the Institute of Education, University of London. She is the author of *Spaced out: policy, difference and the challenge of inclusive education* (Kluwer 2003). Her most recent book, co-edited with Michele Moore, *Action research for inclusive education: changing places, changing practices, changing minds* (RoutledgeFalmer 2004) examines exclusion and inclusion in education using theoretical/ practitioner based research.

Jo-Anne Dillabough is an Associate Professor in the Department of Educational Studies at the University of British Columbia. With Madeleine Arnot, she co-edited *Challenging democracy: international perspectives on gender, education and citizenship* (RoutledgeFalmer 2000). She is also co-editor (with A. Halsey, H. Lauder and P. Brown) of *Education, globalisation and social change* (Oxford University Press 2006). Recent areas of research include feminist analyses of youth poverty and the study of youth culture and social exclusion in urban centres.

Bill Green is a Professor of Education in the Faculty of Education at Charles Sturt University, Bathurst, Australia. His principal research interests are curriculum inquiry and literacy studies, curriculum history, particularly the history and politics of English teaching and the English subjects, and doctoral research education. He is interested in the cultural and educational implications of the 'modernism-postmodernism' debate and in the use-value of poststructuralist theory and philosophy.

Kalervo Gulson is a lecturer in the Faculty of Education, Charles Sturt University, Bathurst, Australia. His research is concerned with employing spatial theories to explore the interplay of urban change, education policy, and identity/subjectivity. Recent papers have appeared in *Discourse: Studies in the Cultural Politics of Education* and the *Journal of Education Policy.*

Jacqueline Kennelly is currently a doctoral student in the Department of Educational Studies at the University of British Columbia, Canada. Her current research focuses on youth cultures, citizenship, social movements, and social exclusion. Her work has been published in such journals as

Canadian Woman Studies, *Canadian Journal of Education*, and the *Journal of Curriculum Theorizing*.

Will Letts is a senior lecturer in the Faculty of Education at Charles Sturt University, Bathurst, Australia. His research interests include the cultural studies of science and science education, with a focus on sexuality, gender, indigenous knowledges, place, and space. With James Sears, he edited *Queering elementary education* (Peter Lang 1999) and he has published articles in *Gender and Education*, *Anthropology and Education Quarterly* and the *Journal for Research in Science Teaching*.

Bob Lingard holds the Andrew Bell Chair of Education at the University of Edinburgh. He has published extensively in the sociology of education and on educational policy. He is editor with Jenny Ozga of the *RoutledgeFalmer reader in educational policy and politics* (2006) and co-author with Debra Hayes, Martin Mills, and Pam Christie of *Leading learning: making hope practical in schools* (Open University Press 2003). He is also an editor of the journal, *Discourse: Studies in the Cultural Politics of Education*.

Pauline Lipman is Professor of Education at the University of Chicago. Her research focuses on the social context of education, and race and class in schools. She is the author of *Race, class, and power in school restructuring* (SUNY Press 1998), *High stakes education: inequality, globalization, and urban school reform* (Routledge 2004), and her articles have appeared in journals such as *American Educational Research Journal*, *Cultural Logic*, *Race, Ethnicity and Education*, *Urban Review*, and *Globalization, Societies, and Education*.

Mario Novelli is a lecturer in International Development at the University of Amsterdam. His research explores the relationship between globalization, education, and development in the 'south' and the role of education in social movement strategies. Recent articles have appeared in such journals as *Societies and Education*. He also has chapters in S. Ludlam et al.'s, *Labour, the state, social movements and the challenge of neo-liberal globalisation* (Manchester University Press 2006) and C. Van den Anker's, *The essentials of human rights* (Hodder Arnold 2005).

Fazal Rizvi has been a Professor in the Department of Educational Policy Studies at the University of Illinois at Urbana-Champaign since 2001. From 1993 to 2000, he edited *Discourse: studies in the cultural politics of education*. His current research deals with issues of globalization and educational policy, student mobility and the internationalization of higher education. His recent books include *Globalization, the OECD and education policy making* (Henry et al., Pergamon 2001) and *Education policy and the politics of change* (Taylor et al., Routledge 1997).

Susan Robertson is a Professor of Sociology of Education in the Graduate School of Education, University of Bristol. She is founding co-editor of the journal *Globalisation, Societies and Education*. Her recent work includes analyses of the various regional agreements such as the EU and their implications for education, as well as analyses of new educational spaces that are being generated as part of state's knowledge economy strategies. She is the author of *A class act: changing teachers' work, the state, and globalization* (Routledge 2000).

Mona Shrestha is a postdoctoral research fellow at the University of Western Sydney. She researches community-based education, migrants, and gender, and cooperative conflict and has written on non-formal education, gender and intergenerational analysis in education, gender and religion, low-income housing, and social eviction. Her publications have appeared in *The Handbook on Educational Research in the Asia Pacific Region* (Kluwer 2003) and the *International Journal of Learning*.

Michael Singh is a Professor of Education at the University of Western Sydney. His research examines the socio-economic implications of globalization on educational pedagogies, the changing global/local movements of people, and the politics of multicultural education. His books include *Globalising education: policies, pedagogies and politics* (with Michael Apple and Jane Kenway) (Peter Lang 2005) and *Adult education@ 21st Century* (with Peter Kell and Sue Shore) (Peter Lang 2004).

Colin Symes is a lecturer in the Australian Centre for Educational Studies at Macquarie University. He is an editor of the *Asia-Pacific Journal of Teacher Education*. His most recent book *Setting the record straight: a material history of classical music* (Wesleyan University Press 2004), was a recipient in 2005 of a Deems Taylor Award from the American Society of Composers, Authors and Publishers. Recent articles have appeared in *Teaching in Higher Education*, *British Journal of Music Education*, *Popular Music*, and *Discourse: studies in the cultural politics of education*.

Chris Taylor is a lecturer in the School of Social Sciences, Cardiff University. His research has focused on the impact of education marketization and school diversity. He has written two books: *Schools, Markets and Choice Policies* (RoutlegeFalmer 2003) and *Geography of the 'New' Education Market* (Ashgate 2002) and has published in the *Oxford Review of Education*, *Environment and Planning A*, *Area*, and the *Journal of Education Policy*.

Claudia Hanson Thiem is a doctoral student in Geography at the University of Wisconsin-Madison. Her dissertation explores the processes of neoliberal state restructuring through case studies of privatization in the US compulsory education sector. She is also reviewing research on the 'geographies of education' as part of an effort to revitalize geographical work in this area.

Pat Thomson is Professor of Education at the University of Nottingham, and adjunct Professor of Education at the University of South Australia.

Her research interests include social justice and policy. Her book *Schooling the rustbelt kids: making the difference in changing times* (Allen and Unwin 2002) employed concepts from human geography, anthropology, and sociology to examine schools in disadvantaged communities. Forthcoming publications include *Romancing the principal: popular pedagogies of school administration* (Peter Lang), and (with Barbara Kamler) *Helping students write: pedagogies for doctoral supervision* (RoutledgeFalmer).

Eugenia Wang is a doctoral student in Geography at the University of British Columbia, Vancouver. Her current research traces the linkages between the production of separated refugee youth in Canada and the state discourses that are mobilized to justify the inclusion or exclusion of youth within a liberal humanitarian framework. She has published in the *Journal of Curriculum Theorizing*.

Index

Page numbers in italics refer to figures or tables.